James Casey

Exploring Curvature

James Casey

Exploring Curvature

With 141 Illustrations

Professor James Casey
Department of Mechanical Engineering
University of California, Berkeley
Berkeley, CA 94720-1740
USA
E-Mail: jcasey@euler.berkeley.edu

Vieweg ist a subsidiary company of the Bertelsmann Professional Information.

Printing and binding: Hubert & Co., Göttingen
Printed on acid-free paper
Printed in Germany

ISBN 3-528-06475-7

FOR IRENA, DAVID, DANIEL, AND EOGHAN

Preface

> . . . one should not be too ready to erect a wall of separation between nature and the human mind.
>
> *d'Alembert* [Dugas (1955)]

It is possible to present mathematics in a purely formal way, that is to say, without any reference to the physical world. Indeed, in the more advanced parts of abstract algebra and mathematical logic, one can proceed only in this manner. In other parts of mathematics, especially in Euclidean geometry, calculus, differential equations, and surface geometry, intimate connections exist between the mathematical ideas and physical things. In such cases, a deeper (and sometimes quicker) understanding can be gained by taking advantage of these connections. I am not, of course, suggesting that one should appeal to physical intuition whenever one gets stuck in a mathematical proof: in proofs, there is no substitute for rigor. Rather, the connections with physical reality should be made either to motivate mathematical assumptions, or to introduce questions out of which theorems arise, or to illustrate the results of an analysis. Such interconnections are especially important in the teaching of mathematics to science and engineering students. But, mathematics students too have much to gain by familiarizing themselves with the interconnections between ideas and real things.

The present book explores the geometry of curves and surfaces in a physical way. All around us, in sculpture and architecture, in engineering and technology, in animals and in plants, we encounter geometrical objects, especially curved surfaces, that fascinate the eye and nourish the imagination. Such objects provide an inexhaustible source of geometrical phenomena.

It was not until the end of the 18th century that mathematicians had developed sufficiently powerful analytical tools to describe the properties of curved surfaces. The subject came of age in the early 19th century at the hands of the great mathematician Gauss, who was the first to recognize that the essence of surface geometry lies in the *intrinsic curvature* of surfaces. This is a measure of their non-flatness, or non-Euclideanness. The subsequent exploration of the notion of curvature by Riemann in the mid 19th century had a profound influence on the development of modern geometry. Even more striking perhaps was its eventual central role in Einstein's theory of general relativity. However, as mentioned before, a plentiful supply of curved surfaces lies even closer at hand. Vases, watermelons, liquid-detergent bottles, *etc.*, all possess surfaces with interesting geometrical features. To study these surfaces analytically requires advanced mathematical tools, and yet, their essential geometrical properties can be explored physically in a straightforward manner. In the present book, a series of simple experiments is described, by means of which the reader can develop an understanding of the geometrical phenomena exhibited by curves and surfaces. In addition, the fundamental geometrical concepts are explained rather thoroughly, but in a way that links them to the experiments. All too often, students see geometry only in its polished format of definitions, axioms, theorems and proofs, and while they should be impressed by its austere beauty, they must also be made to realize that geometry was not, and is not, *created* in accordance with such a rigid plan. In other words, the mode of discovery is very different from the mode of final presentation. There are good reasons for this—having to do mainly with logical rigor—but nevertheless, if mathematics is to be appreciated as a creative endeavor, one must also try to approach its contents from a more organic viewpoint. In the presentation that follows, the geometry of curves and surfaces is unveiled gradually through the experiments and is carefully elucidated in the accompanying discussions of the concepts. An open-minded reader progressing at a leisurely pace will soon experience the excitement of mathematical discovery, and it is hoped that he or she will be inspired to study the formal mathematical theory in a university course on differential geometry.

Using experiments as an aid to teaching geometry is not new. An early advocate of this approach was William George Spencer (father of the philosopher Herbert Spencer) who wrote a highly original little book entitled *Inventional Geometry*, the American edition of which was published in 1876. In the modern literature, there is a sizeable body of work on the intuitional presentation of elementary geometry. A recent valuable addition is D. W. Henderson's *Experiencing Geometry on Plane and Sphere*.

While most of the material in the text is self-contained, some of the derivations presume a knowledge of calculus.

The book also contains some historical and biographical material. Personally, I feel strongly that mathematics is an integral part of human culture and should be presented in that way whenever the opportunity arises. Indeed, many technical and non-technical aspects of mathematics can only be fully appreciated from a study of the cultural – and especially the philosophical – context within which mathematicians work. It is also interesting to find out how the creators of this beautiful subject felt about it and how they incorporated it into their daily lives.

For source material, I have depended upon a wide variety of excellent references in geometry, history of mathematics, and biography. These are included in the Bibliography, but it is only proper to draw special attention here to the lucid writings of the contemporary mathematicians S.-S. Chern, A.D. Aleksandrov, and P.S. Aleksandrov, to the historical work of the mathematicians A. Seidenberg and B.L. van der Waerden, and to the insightful, scholarly essays in the *Dictionary of Scientific Biography*. Valuable surveys of recent work in geometry can be found in the October 1990 issue of *The American Mathematical Monthly*.

A number of people have helped me to bring this book into existence. It is with pleasure that I thank Jean Pedersen of Santa Clara University for her generous advice and comments on an embryonic version of the manuscript. Andy diSessa of the University of California at Berkeley also made useful suggestions at that stage. Herb Clemens of The University of Utah kindly invited me to present some of the experiments at the 1992 Summer Geometry Institute in Utah. I thank John Sullivan of the Geometry Center at the University of Minnesota for sharing with me material which he had prepared on curvature of polyhedra. It was a joy also to work through some of the subject matter at Berkeley High School and Piedmont High School, with Heidi Boley and her students and with Doyle O'Regan and his students, respectively. The comments of Laura Risk, Deepak Nath, and Eveline Baesu on various chapters are much appreciated. I am grateful to MaryAnne Peters, Tom Dambly, Bonita Korpi, and JoAnn Nerenberg for their efforts in typing various parts of the manuscript. Also, the help of my colleague, Panos Papadopoulos, in creating a Latex file of the book is much appreciated. The figures were expertly drawn by Cynthia Borcena-Jones, using Adobe Illustrator. I thank Herb Ranharter for taking the photographs. The portraits of Gauss, Riemann, and Levi-Civita were supplied by Konrad Jacobs of Universität Erlangen Nürnberg. I am especially indebted to my editor Ulrike Schmickler-Hirzebruch of Verlag Vieweg, for her advice and patience, and for guiding the project to completion.

James Casey
July 1996

Table of Contents

To the Reader: How to Use this Book

This book is not meant to be just read. It is structured around a series of experiments, which I encourage you to perform carefully. Before carrying out an experiment, you should study the relevant portion of the text, form your own questions and conjectures, and ask yourself how you might check these out experimentally. As you do each experiment, watch out for other geometrical phenomena besides the one you are primarily concerned with. An experiment is a dialogue with Nature, and by the end of it some knowledge is usually revealed. My main objective is to encourage you to raise and explore mathematical questions as soon as you have grasped the requisite basic ideas. You will explore geometrical questions in much the same way as you do experimental physics. You are probably not used to approaching mathematics in this way: mathematicians prefer rigorous logical proofs and difficult calculations as ways of addressing questions. Nevertheless, you will gain a valuable understanding of geometrical phenomena by doing the experiments and thinking deeply about the concepts that are involved. My main intent is to get you to look in a new way at the curves and surfaces that lie all around us, to *do* things to them, to dissect and deform them, to manufacture new surfaces – until you have discovered their geometrical secrets.

For the most part, all necessary concepts are explained in detail, but a number of important derivations require a knowledge of calculus. Many excellent textbooks on calculus are available, but for its clarity and depth my favorite is Courant (1934) – listed in the Bibliography. On a first reading of the present book, it is probably advisable to skip any details that you cannot follow, in order to get on with the experiments.

The symbol □ is used to indicate the end of an experiment.

List of Experiments

1

The Evolution of Geometry

Geometry is concerned primarily with the spatial properties of objects and with the endless stream of abstract generalizations to which these properties give rise. Modern geometry is an extremely active field of research by pure and applied mathematicians, and it also has significant applications in physics and engineering. In the present book, we will explore in a physical manner the geometrical properties of curves and surfaces, and will discuss in detail concepts that are essential to modern geometry.

By way of background, and also to give the reader a glimpse of the rich history of the subject, in the present chapter I summarize the key geometrical ideas that have become part of our historical heritage.

The Beginnings

Two great traditions can be easily recognized in the development of mathematics: the constructive or geometric, and the calculational or algebraic. The first arises from spatial intuitions, the second from numerical perceptions. Recent studies in the history of mathematics have indicated that the two traditions stem from a single source.[1] The original body of mathematics, invented at least 4500 years ago, was in the possession of Indo-Europeans living in Neolithic Europe. These people may themselves have been the inventors of mathematics, or they may have inherited it from some still older group. In any case, the mathematics which they knew became the basis of all later ancient mathematical developments, in India, China, Babylonia, Egypt, and Greece, *etc.*

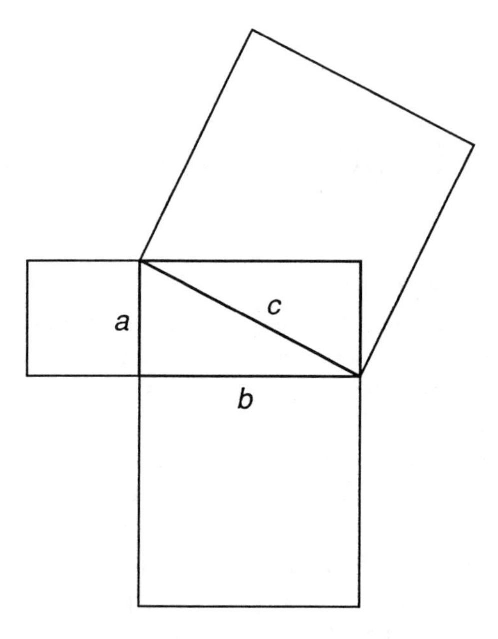

Figure 1 The "Theorem of Pythagoras"

A central element in the mathematics of the neolithic period was a result that is universally known as the Theorem of Pythagoras. Traditionally ascribed to Pythagoras, a Greek mathematician, philosopher, and mystic, who lived in the sixth century B.C., this theorem had an extensive history prior to its prominent occurrence in Greek geometry. The result was known originally in both its constructive and calculational aspects. Thus, on the one hand it was asserted that the square constructed on the diagonal of a rectangle has an area equal to the sum of the areas of the squares constructed on two adjacent sides of the rectangle (Fig. 1). This is a geometrical statement and corresponds to the version of Pythagoras's Theorem that was known in ancient India. On the other hand, the length c of the diagonal was said to be given by

$$c = \sqrt{a^2 + b^2} . \qquad (1.1)$$

Clearly, the calculation indicated in Equation (1.1) can be performed without any knowledge of the squares constructed in Fig. 1. Evidence of the distinction between the constructive and calculational traditions can still be seen in the usual division of high school mathematics into geometry and algebra. The distinction is an artificial one.

The Indian texts in which Pythagoras's Theorem appears are called the *Śulvasūtras* and are concerned with the exact construction of altars used in sacred rituals. Altars of various shapes were required, some simple, some very elaborate. Two altars were considered to be ritually equivalent if their areas were equal. Out of this theological idea springs a host of geometrical problems. For instance, how does one construct a circular altar that is equivalent to a given square one?

The word " Śulvasūtras " may be translated as "Rules of the Cord", which refers to the practice among the Indian ritualists of employing cords, along with poles, to construct the figures of the altars. This elementary method of construction is quite accurate and is known to have been used in ancient Egypt by temple designers, and probably also by the builders of megalithic henges in Great Britain about 3000 B.C.

In our first experiment, we replicate on a small scale two "peg and cord" constructions of a right angle:

Experiment 1 (Constructing a right angle): Take a flat board of soft wood, measuring 30 cm by 30 cm approximately, and affix a sheet of cardboard to it using small nails. Then, place a sheet of paper on top of the cardboard and secure it with thumbtacks. For pegs, small nails or push-pins can be used, and for cord, any non-stretchable type of sewing thread works fine.

(**a**) To erect a perpendicular at a given point E on a given line drawn on the sheet of paper (Fig. 2), proceed as follows:

(1) Take a thread of any convenient length l and tie loops at both ends of it;

(2) Fold the thread to find its midpoint C, and mark it there;

(3) Let E be the point on a given line at which a perpendicular is to be

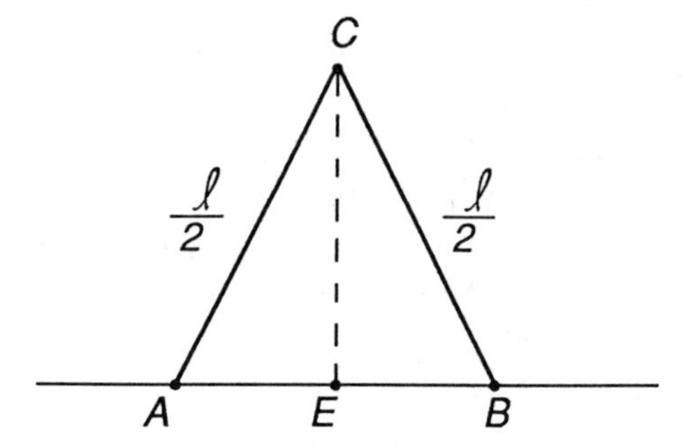

Figure 2 Ancient construction of a perpendicular

erected. Using the thread, take any two points A and B that lie on the line and are equidistant from E, but less than a distance l from one another. Hammer nails into the board at A and B;

(4) Place the loops over nails A and B;

(5) Pull the thread taut by a push-pin held against C. Hammer in a nail at the point where C ends up. The line CE is perpendicular to AB.

(**b**) Here is an alternative method, also dating back to the *Śulvasūtras:*

(1) Place nails A and B in the board a distance of 4 units (*e.g.*, 12 cm) apart (Fig. 3);

(2) Take a thread (with end loops) of length 8 units and mark a point C on it which is 3 units from one end;

(3) Loop the thread over the nails, and holding a push-pin against it at C, stretch it taut. Drive a nail at the point where C ends up. Then the angle ABC is a right angle.

Use a protractor to assess the accuracy of your constructions. □

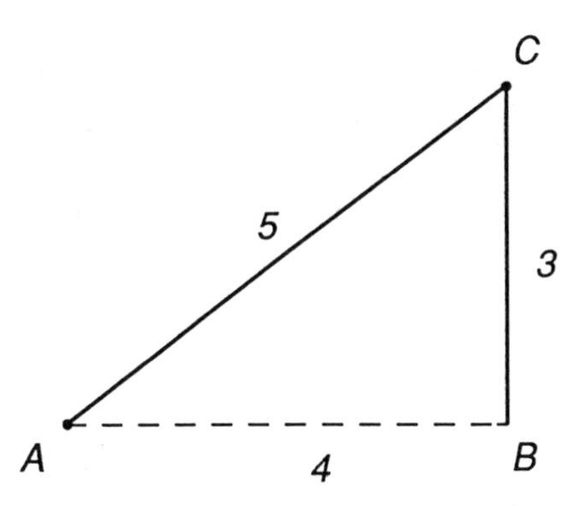

Figure 3 Another ancient construction of a right angle

The integers (3, 4, 5) appearing in Part (b) of Experiment 1 satisfy Equation (1.1) and form a *Pythagorean Triple*, and the corresponding triangle is called a *Pythagorean Triangle*, Several other Pythagorean Triples and Triangles were known in ancient civilizations. For example, the Indians had (5, 12, 13), (8, 15, 17), and (12, 35, 37) and others, the Chinese had a rule for computing Pythagorean Triplets, and a clay tablet (designated as Plimpton 322) and inscribed in Babylonia during the reign of Hammurabi (*c.* 1700 B.C.), contains a long table of Pythagorean Triplets.

Among the Egyptians, the tradition of rope-stretching can be traced back at least to the period 2800 - 2500 B.C., during the Old Kingdom. The Greek mathematicians thought highly of the Egyptian priest-geometers, whom they called *harpedonaptai* (= rope-stretchers). Some of the Greeks were under the impression that Egyptian geometry had its origin in land-surveying, noting that the annual inundation of the Nile altered the boundaries of fields, which had to be newly measured when the water subsided. However, in Plato's *Phaedrus*, Socrates says that the god Thoth (whom the Egyptians associated with writing, measuring, and calculating) is attributed with the invention of geometry (as well as arithmetic and astronomy). Aristotle also does not suppose a practical origin for geometry, but simply suggests that the Egyptian priests, being a leisured class, had ample time for such activities.

It is commonly held that proofs did not enter mathematics until the Greeks realized the necessity for them. This is not correct, as there are actually some proofs in Indian and Chinese mathematics, and there is every indication that one or other proof of Pythagoras's Theorem lay within the reach of the ancient mathematicians. As the twentieth-century mathematician and historian van der Waerden puts the matter:[2]

> *I am convinced that the excellent neolithic mathematician who discovered the Theorem of Pythagoras had a proof of the theorem . . .*

As mentioned before, the Babylonians of about 1700 B.C. knew about Pythagorean Triplets. The mathematics of the Babylonians (which has been preserved on clay tablets inscribed in cuneiform script), is essentially algebraic in character: the emphasis is always on calculation rather than on construction. It epitomizes the second of the two great traditions in mathematics.

At some remote epoch, the two traditions separated and gradually developed distinct styles and attitudes. For example, in algebra one traditionally manipulates literal symbols according to certain rules and learns to solve equations, whereas in geometry one encounters postulates, theorems, and proofs. It has been argued that the reason for the initial separation of the traditions was that the ritualists were unable to find a geometrical solution to the problem of converting a circle into a square of equal area. The solution that they had involved a numerical *approximation* to $\sqrt{2}$. One branch of ancient mathematics expanded on arithmetical methods, leading eventually to Babylonian mathematics, while the other confined itself to exact geometrical methods, and eventually emerged as Greek geometry.

While it can no longer be maintained that exact mathematical knowledge *originated* in Greece during the latter part of the first millennium B.C., nevertheless, the Greek philosophers, mathematicians, and astronomers deserve our admiration for the profundity of their contributions to mathematics. In Classical Greece, mathematical thought flourished and rapidly attained a level of rigor which was surpassed only towards the end of the 19th century. Greek views on the nature and role of mathematics have had a lasting influence. Furthermore, the Greeks imparted upon mathematics an abstract and deductive organization that has remained a hallmark of the subject.

The inception of philosophy and geometry in Greece has been credited to a single individual, Thales of Miletus, who lived from approximately 625 B.C. to approximately 547 B.C. According to later Greek sources, Thales learned geometry from the Egyptian priests, and then proceeded to make discoveries of his own. A number of elementary geometrical statements are associated with his name, among which we mention the following:[3]

(1) A circle is bisected by its diameter;

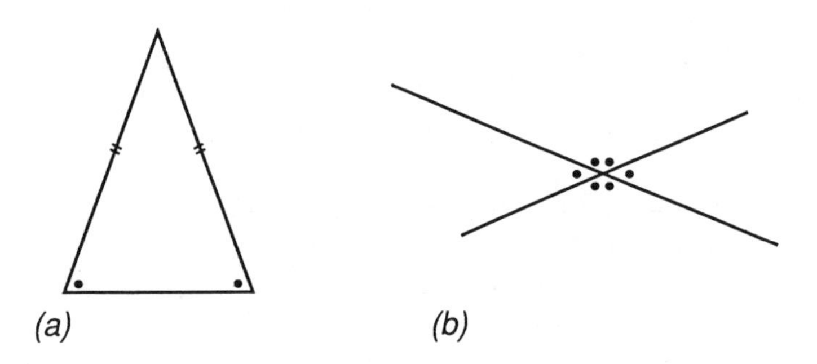

Figure 4 Some results attributed to Thales

(2) The angles at the base of any isosceles triangle are equal (Fig. 4a);

(3) When two straight lines intersect, the opposite angles are equal (Fig. 4b).

For at least some of the statements attributed to Thales, the Greek historians of mathematics say that he was in possession of a demonstration. These proofs have not been handed down, but there is a suggestion that Thales's method involved folding the figure. Such a procedure would not have satisfied later Greek mathematicians, and would therefore have been discarded from official mathematics. Nevertheless, Thales's contribution to geometry is significant: he isolated certain elementary geometrical statements, which would re-appear as important building bricks in the magnificent edifice which Greek geometry was to become; moreover, he appreciated the necessity for proofs. Thales is also regarded as the first Greek philosopher. In his speculations, which were concerned chiefly with the nature of the universe, he attempted to identify fundamental universal principles which would explain the workings of the world in physical terms, *i.e.*, without any appeal to supernatural forces. This was a crucial step in the evolution of human thinking.

In the earliest Greek philosophy, we find very powerful statements about how the world works, but we do not yet see mathematics playing any role in the theory. This key element was injected into Greek philosophy by Pythagoras and his followers, whose fundamental contributions to Western civilization are difficult to overestimate.

The island of Samos was home of a flourishing Greek community in the second part of the 6th century B.C. It was here that the philosopher, mystic, and mathematician Pythagoras was born (about 560 B.C.) and spent his early years. He is said to have then studied with Thales, who advised him to go to Egypt for further instruction.[4] It is also reported that he studied in Phoenicia and Babylon.

After returning to Samos from his long travels, Pythagoras soon grew dissatisfied with its ruler Polycrates, and emigrated in 530 (or perhaps 520) B.C. to the city of Crotona (modern Crotone) in southern Italy. There, he founded a community of men and women devoted to mysticism, philosophy, and mathematics. The sect played an active role in the political life of Crotona, but this involvement eventually led them into trouble. After a number of his followers were killed, Pythagoras departed for Metapontum (near present-day Metaponto) where he is reported to have died (in 480 B.C. or thereabouts). Branches of the Pythagoreans lived on for some 200 years longer, and continued to exert a powerful influence both on Greek thinking and on Greek religion.

Neither Pythagoras himself, nor any of his immediate successors, left a written account of his teachings; evidently, as was the case in many ancient cultures, their mode of transmission of wisdom was exclusively that of oral communication. We depend upon authors of a later period for the information we have on the Pythagoreans, and it is often difficult to separate fact from legend in such circumstances.

The Pythagorean philosophy was a tightly woven blend of metaphysics, mathematics, mysticism, and asceticism.[5] Each part of it drew sustenance from the rest. In particular, mathematics received vital nourishment by being an integral part of the Pythagorean organism, and as its importance to Pythagorean philosophy grew, so also did the efforts of the Pythagoreans to increase and deepen their understanding of mathematics. Proclus, the last important Greek philosopher (and mathematician), living almost a millennium after the inception of the Pythagorean movement, wrote:[6]

> . . . *Pythagoras transformed this study* [geometry] *into the form of a liberal education, examining its principles from the beginning and tracking theorems immaterially and intellectually;*

Pythagoras believed in an orderly arrangement of the universe. For him, the universe has a living and breathing unity held together by mathematical relations. At the center of his religious teachings were the doctrines that the soul is immortal, that there is a kinship between all living things, and that upon death, the soul can migrate to another person, or animal, just being born. As part of their religion, the Pythagoreans were bound by a series of prohibitions, such as abstinence from meat (and beans too!), and were also forbidden to divulge the teachings of the Master to outsiders. By meditating on the mysteries of the universe, by contemplating its mathematical laws, and by following the rules and practices of the sect, the Pythagoreans hoped to purify their souls and achieve salvation.

Pythagoras held that at the very deepest level the orderliness of the universe was mathematical. As Aristotle recounts:[7]

> *. . . the Pythagoreans, as they are called, devoted themselves to mathematics; they were the first to advance that science, and having been brought up in it they thought that its principles were the principles of all existing things. Since of these principles numbers are by nature the first, and in numbers – more than in fire and earth and water – they seemed to see many resemblances to the things that exist and come into being . . . since, again, they saw that the attributes and the ratios of the musical scales were expressible in numbers; since, then, all other things seemed in their whole nature to be modelled after numbers, and numbers seemed to be first things in the whole of nature, they supposed the elements of numbers to be the elements of all things, and the whole heaven to be a musical scale and a number. And all the properties of numbers and scales which they could show to agree with the attributes and parts and the whole arrangement of the heavens, they collected and fitted into their scheme; and if there was a gap anywhere, they readily made additions so as to make their whole theory coherent.*

What we see emerging here is an intellectual attitude to nature which was to become the guiding principle of natural philosophy, and which remains an essential characteristic of science even to the present day. Indeed, all physical scientists believe that nature is explicable in mathematical terms, and that it is through mathematics that the beauty and harmony of the world is most clearly expressed. Evidently, Pythagoras would go even further than this and claim that things are *made* out of numbers, which sounds strange to modern ears. The faith of the Pythagoreans in the power of mathematics as a governing principle of all natural phenomena had no bounds. They were completely enthralled by the correspondences which they were able to draw between the patterns that they observed in their mathematics and the order that they perceived in the universe.

The kingpin in Pythagoras's theory of the orderliness of the universe was the correspondence which he discovered between consonant tones in music and small (whole) numbers. Thus, he observed that if two lengths of vibrating string produce tones one octave apart, then the ratio of the lengths is 1:2 (the shorter piece producing the higher tone). Likewise, the musical intervals called fifths and fourths correspond respectively to ratios 2:3 and 3:4. Pythagoras generalized this correspondence principle to all of nature: relations between numbers accounted for the harmonious arrangement and operation of the universe. This profound new viewpoint was certainly the most important philosophical contribution of the Pythagoreans.

Let us now examine some of the specific contributions that the Pythagoreans made to mathematics.

From what was said earlier in this chapter, it should be clear that strong algebraic and geometric traditions of mathematics existed prior to the flowering of philosophical thought in Greece in the 6th century B.C. Pythagorean mathematics shows the influence of both traditions and tried to treat them in a cohesive way. For the Pythagoreans, of course, numbers—meaning the positive integers—were the primitive ingredients of all mathematics. They were fascinated with the patterns which they discovered in numbers, and their findings gave a tremendous impetus to number theory, and to number mysticism as well. They were struck by the distinction between odd and even numbers, and for some reason regarded the former as male and the latter as female. They knew the concept of a prime number, which has proven to be fundamental to number theory. And, of course, they studied Pythagorean Triples. (But, they also had some less fruitful ideas about numbers.)

In geometry, Pythagoras and his followers made much progress, increasing its body of results, providing proofs, and imparting an organi-

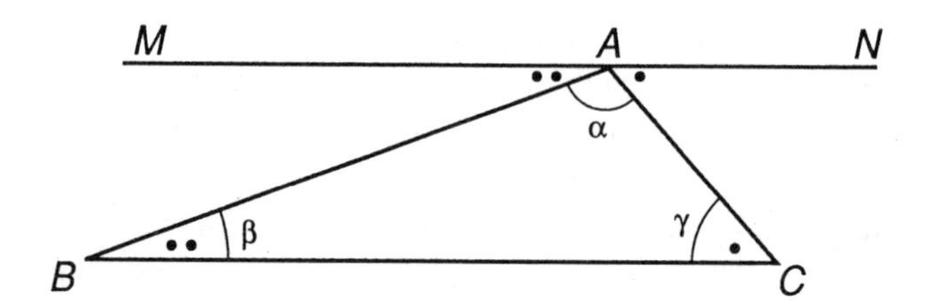

Figure 5 Theorem on angle sum of a triangle

zation to the subject. The following central theorem of Greek geometry is attributed to the Pythagoreans:[8]

> The sum of the interior angles of any triangle
> is equal to two right angles.

Thus, in any triangle *ABC* (Fig. 5), we have

$$\alpha + \beta + \gamma = 180^\circ . \tag{1.2}$$

The Pythagoreans gave a nice proof of this result, which may be recounted as follows: Through *A* draw the straight line *MN* parallel to *BC*. Then, the angle *NAC* must be equal to *ACB* ($= \gamma$), and the angle *MAB* must be equal to *ABC* ($= \beta$). But, the three marked angles at the vertex *A* add up to two right angles (or 180°), since *MAN* is a straight line. Therefore *ABC* plus *ACB* plus *BAC* must also add up to 180°.

Next, we meet once again the famous theorem which bears the name of Pythagoras. The theorem is certainly a magnificent result, and Pythagoras surely realized its power. He is said to have sacrificed an ox upon (allegedly) discovering the theorem. Whatever the basis of this story, we may conclude from it that the importance of the theorem was strongly felt in Greek times. As we shall see presently, the theorem is the crowning result in Book I of Euclid's *Elements*, but it is relevant to mention

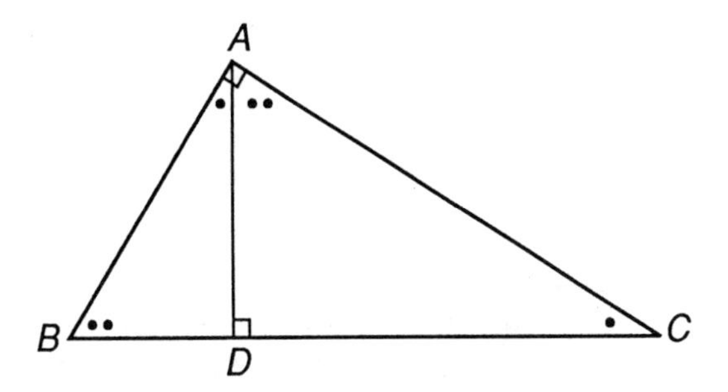

Figure 6 Similarity proof of The Theorem of Pythagoras

here a proof which is possibly due to Pythagoras himself: Let ABC be a triangle having a right angle at A (Fig. 6). Drop a perpendicular from A to meet BC at D. Examining various angles in the figure, we see that each of the triangles ABD and ADC is similar to the triangle ABC (and that they are similar to one another).[9] From the similarity of ABD and ABC, we obtain

$$BA : BC = BD : BA . \tag{1.3}$$

Likewise, from the similarity of ADC and ABC, we have

$$AC : BC = DC : AC . \tag{1.4}$$

Hence,

$$BA^2 = BC \cdot BD \tag{1.5}$$

and

$$AC^2 = BC \cdot DC . \tag{1.6}$$

Adding Equations (1.5) and (1.6), and noting that

$$BC \cdot BD + BC \cdot DC = BC \cdot BC \ , \tag{1.7}$$

we conclude that

$$BA^2 + AC^2 = BC^2 \ . \tag{1.8}$$

which establishes the desired result.

As you will have gathered, a central dogma in the Pythagorean philosophy is that "number rules the universe", as it is often succinctly put. However, since by *number*, the Pythagoreans meant only *positive integer*, this dogma resulted in a serious intellectual problem for them. In the context of geometry, this dogma would imply that every two of line segments AB and EF possess a common measure (or, are *commensurable*), *i.e.,* that there exists some line segment PQ such that

$$AB : PQ = n : 1 \tag{1.9}$$

and

$$EF : PQ = m : 1 \ , \tag{1.10}$$

where n, m are positive integers. It would then follow that the ratio of the segments EF and AB could be expressed as the ratio of two integers:

$$EF : AB = m : n \ . \tag{1.11}$$

At some stage in their geometrical meditations (probably during the latter part of the 5th century B.C.), the Pythagoreans realized that the property (1.11) cannot possibly hold for all line segments; in particular, if BC is the diagonal of a square having a side AB (Fig. 7), then BC and AB possess no common measure. This may be argued as follows: Suppose that BC and AB are commensurable. Then, we could find integers such that Equations of the type (1.9) and (1.10) would hold for AB and BC, respectively. If these integers had a common factor greater than 1, we could divide it out. Hence, we would have

$$BC : AB = m : n \ , \tag{1.12}$$

where $m : n$ has been reduced to its lowest terms. Hence,

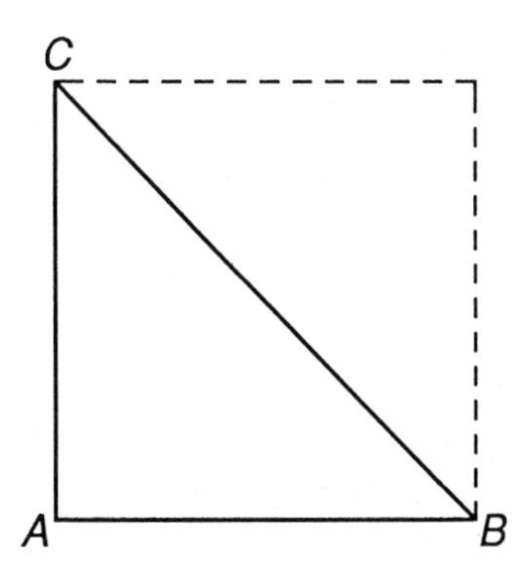

Figure 7 Incommensurability of diagonal and side of a square

$$BC^2 : AB^2 = m^2 : n^2 . \tag{1.13}$$

But, applying the result (1.8) to the isosceles right-angled triangle in Fig. 7, we have

$$BC^2 : AB^2 = 2 : 1 . \tag{1.14}$$

Consequently,

$$m^2 = 2\,n^2 . \tag{1.15}$$

It follows from Equation (1.15) that m^2 must be an even number, and hence m must also be even (because an odd number multiplied by itself is odd). So, m can be written as

$$m = 2k \tag{1.16}$$

for some positive integer k. From Equations (1.16) and (1.15), we deduce that

$$n^2 = 2k^2 , \tag{1.17}$$

and hence n^2 is even, and n is even. Consequently, both m and n are even and therefore have a common factor 2. This contradicts our previous statement that m and n have no common factor. Thus, the assumption that BC and AB are *commensurable* leads to a contradiction. Therefore, they must be incommensurable.

The foregoing argument, which is probably due to the Pythagoreans themselves,[10] forced them to accept that number (as they understood it) does not rule even line segments. This realization must have had a devastating effect on their philosophy. We are told that the first Pythagorean to reveal the discovery of incommensurability to outsiders perished in a shipwreck! For Pythagorean geometry, the discovery of incommensurability was also ravaging: the Pythagorean theory of similarity contained an implicit assumption that every two line segments are commensurable. In particular, the Pythagoreans could no longer have any confidence in their similarity proof of Pythagoras's Theorem.

The crisis experienced by the Pythagoreans is the first example I know of in which a theory about nature is shown to be inadequate by means of a *thought-experiment, i.e.,* by applying their theory that number rules the universe to an idealized problem (in the present case, a geometrical one). Theoretical physics, as we know it, habitually progresses to new stages by the formulation and resolution of paradoxes arising from thought-experiments. Mathematics, too, gains in the long run from logical crises: they have the beneficial effect of encouraging mathematicians to state their explicit assumptions carefully and to examine their arguments for deeply hidden implicit assumptions, but additionally, they sometimes lead to a completely new understanding of a concept. However, in the case of the discovery of incommensurability, the immediate effect on geometry was a debilitating one. It caused the Pythagoreans to divorce geometry from numbers. Instead of enlarging their concept of number to include irrational numbers, they expunged number from geometry. This led to an awkward and laborious approach to geometry, an approach which, however, the Greek mathematicians mastered and brought to perfection.

Because of their seminal importance in the history of the mathematical sciences, I have devoted considerable attention to the Pythagoreans. It should be added that during the 5th century B.C., mathematics and natural philosophy were being pursued by other Greek scholars as well. Noteworthy among these were: Democritus of Abdera, famous for his ideas on atoms, and author of many mathematical and physical works; Hippocrates of Chios known for his theory of lunes; and Zeno of Elea, whose name is immortalized by the brilliant paradoxes that he invented.[11]

By the end of the 5th century B.C., a considerable amount of mathematical knowledge had been amassed, questions of intrinsic mathematical interest were beginning to be raised, and the need for careful proof was beginning to be appreciated. Mathematics was about to emerge as a discipline in its own right. But, the crystallization of the subject had to await the young, brilliant mathematicians who were to study at Plato's Academy in the 4th century B.C.

Plato's Attitude to Mathematics

Although Plato (427-347 B.C.) was not himself a mathematician, his influence on the early development and eventual stature of the subject has been enormous.[12] Having studied with Socrates (470?-399 B.C.) and having traveled extensively in pursuit of knowledge, Plato established his Academy in Athens sometime between the early 380s and 367 B.C. The brilliant mathematicians Theaetetus (417?-369 B.C.) and Eudoxus (400?-347? B.C.) taught at the Academy, and Aristotle (384-322 B.C.) became a student there at the age of 17, and later remained on for several years as a teacher.

Plato was impressed by the certainty and timelessness of mathematical facts: the number 17, he might say, is beyond doubt a prime, always was, and ever will be. He became convinced of the existence of an independent world of eternal, unchanging ideas, which is accessible through thought; he regarded that world as true reality, and the ordinary world of things and events merely as a cloudy reflection of the world. In his dialogue *The Republic*, he has Socrates say about geometers:[13]

Further you know that they make use of visible figures and argue about them, but in doing so they are not thinking of these figures but of the things which they represent; thus it is the absolute square and the absolute diameter which is the object of their argument, not the diameter which they draw;

The value of mathematics, in Plato's view, is that it draws the mind towards the world of ideas, clarifying it in the process, and purifying man's eternal soul.

Whether or not one subscribes to Plato's views, whether or not one holds that mathematics is invented rather than discovered, it must be admitted that the enthusiasm that Plato expressed for the subject provided an impetus for its systematic development, and was also vital to establishing the deep educational value of mathematical studies. At the Academy itself, not only was mathematics taught as a subject, but also fundamental research on reconstructing, rigorizing, and extending mathematics was pursued. Theaetetus made important contributions to the theory of incommensurable magnitudes and to solid geometry. Eudoxus developed an excellent theory of proportion to replace the defective Pythagorean theory (which applied only to commensurable magnitudes). Another fundamental contribution of his is the *method of exhaustion*, which furnishes a way of approximating areas (and volumes) by a sequence of inscribed polygons (and polyhedra). It is based on the following idea: Let a be a magnitude, arbitrarily large, and let ϵ be a nonzero magnitude, arbitrarily small. Subtract more than $a/2$ from a, and from the remainder subtract more than half of itself. Continuing on in this manner, you can always arrive at a magnitude that is less than ϵ. Thus, for example, while the difference in length between the circumference of a circle and an inscribed polygon can never be reduced exactly to zero, nonetheless, this difference can be systematically made smaller than any pre-assigned small positive number. This is a very fertile notion, and involves the same as the notion of a *limit* of a sequence, an idea which is essential to modern mathematics.[14]

No other name has been so closely associated with geometry as that of Euclid. Few facts about his life have survived, but no mathematical work is better known than his *Elements*.[15] Euclid probably studied under Plato's pupils at the Academy, but spent the main part of his life in the city of Alexandria near the Nile delta. Soon after its foundation in 332 B.C. by Alexander the Great (356-323 B.C.), Alexandria had become the pre-eminent center of culture and learning in the Greek-speaking world. It had a university and a great library. Euclid taught mathematics in Alexandria and wrote his *Elements* (and other works) there.

The *Elements*, in 13 books, is a magnificent logical organization of classical Greek mathematics. Incorporating the newest work by Theaethetus and Eudoxus, Euclid forged the entire subject into a sound logical order, stating definitions, spelling out assumptions that were previously left implicit, proving each theorem step by step, and sequencing the theorems in an elegant fashion, until the whole edifice was completed. After Euclid's example, this manner of presentation became the standard in mathematics. It has many advantages. For instance, the chain of logical connections is clearly laid out and circular arguments and contradictions are more easily avoided. Also, it can result in a very dramatic effect when many novel results are deduced from a small number of accepted (and often innocuous-looking) assumptions: it looks like one is getting something for nothing. The only disadvantage of the method is that it usually does not reveal the manner of discovery of the results, nor the motivations for seeking them in the first place.

The *Elements* contains both plane and solid geometry, and also includes number theory as well as material that is algebraical in character. However, the mode of exposition in the *Elements* is geometrical: only positive integers and their ratios being admitted, the incommensurability of line segments forced Euclid to eschew attaching numbers to line segments; similarly, in his discussion of quadratic equations, he had to use segments (instead of our irrational numbers). Euclid's *Elements* are the epitome of the geometric tradition in mathematics. It should be added, however, that when applied to algebraical material, the Greek geometrical approach is artificial and cumbersome.

Although Euclid's definitions are not always satisfactory, and not all of his assumptions are explicitly stated, nevertheless, the essential geometrical postulates as he knew them are placed at the beginning of Book I of the *Elements*. It is these that characterize what is universally known as *Euclidean geometry*. Thus, for example, regarding straight lines, Euclid assumes that

(a) Between any two points there is a unique line segment; and

(b) Any finite line segment can be extended uniquely and indefinitely from each end to form a straight line.

Assumptions such as (a) and (b) are in agreement with our usual intuitive notions regarding straight lines, and perhaps the reader is wondering why one bothers to write them down at all. The answer is that for logical rigor, one is required to make explicit all assumptions that one uses later in proofs. It is often very difficult to achieve this ideal, but mathematicians cannot be satisfied with a theory until it is perfected in this way. Of course, one always tries to assume as little as possible, and to make the assumptions so simple and clear that any contradiction lurking between them can be detected.

Euclid made one assumption that some later mathematicians felt he ought to have proved. This was his "Parallel Postulate", Postulate 5. Before I quote it, let me recall Definition 23 from Book I of the *Elements*:[16]

> *Parallel* straight lines are straight lines which, being in the same plane and produced indefinitely in both directions, do not meet one another in either direction.

Now, for Postulate 5:

> If a straight line falling on two straight lines make the interior angles on the same side less than two right angles, the two straight lines, if produced indefinitely, meet on that side on which are the angles less than the two right angles.

Thus, referring to Fig. 8, consider two straight line segments, AB and CD, and a straight line EF intersecting both (EF is called a *transversal*). Postulate 5 asserts that if $\alpha + \beta$ is less than 180°, then AB and CD, when produced, will meet in some point G as indicated in the figure. In the first 28 propositions of the *Elements*, Euclid makes no appeal to Postulate 5. Many mathematicians came to regard Postulate 5 as "a blemish on the beauteous body of Geometry". From Euclid's time to the mid 19th century, ingenious attempts were made to prove Postulate 5, but none was successful. It was not until the mathematicians N.I. Lobachevsky (1792-1856) and J. Bolyai (1802-1860), working independently of one another in the 1820s, invented a system of geometry in which Postulate 5 is violated while the rest of Euclid's assumptions remain intact. This was the momentous birth of non-Euclidean geometry (we will say more about the geometry of Lobachevsky and Bolyai in Chapter 21). The logical possibility of non-Euclidean geometry made it clear that Postulate 5 was essential to Euclid's system of geometry, and finally Euclid could receive the credit long overdue to him for recognizing the essential need for Postulate 5 in his system of geometry.

Among those theorems in Book I of the *Elements* that depend on

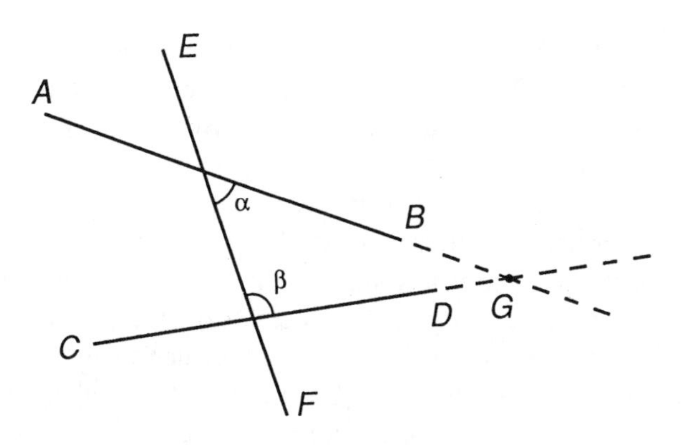

Figure 8 Euclid's Postulate 5

Postulate 5 are two extremely important results which we have already met: (1) the interior angles in any triangle together make up two right angles; and (2) Pythagoras's theorem.

After Euclid came several other fine Greek mathematicians, the most famous among these being Apollonius (262-190? B.C.), who learned mathematics from the pupils of Euclid at Alexandria, and Archimedes (287-212 B.C.) who lived in Syracuse, on the island of Sicily.[17] Apollonius is best known for his eight-volume work on conic sections. Archimedes, who is regarded as the greatest mathematician of antiquity, wrote highly original works in both mathematics and mechanics. For example, using inscribed and circumscribed polygons, he proved that the ratio of the circumference of any circle to its diameter is less than $3\frac{1}{7}$ but greater than $3\frac{10}{71}$ (*i.e.*, $3\frac{1}{7} < \pi < 3\frac{10}{71}$), and he calculated areas and volumes of various figures. As a mark of his genius, it may also be mentioned here that Archimedes realized that the following result, however obvious on intuitive grounds, must be taken as an assumption:[18]

> Of all lines which have the same extremity the straight line is the least.

The reason for the necessity of such an assumption is that in the context of Greek mathematics, there was no satisfactory way of defining the arc length of a curve (a modern definition will be given in Chapter 8).

The sad story of Archimedes's death is well-known: after an eight-month siege by the Roman army of Marcellus, the city of Syracuse finally surrendered; despite Marcellus's orders that the renowned Archimedes should be spared any injury, a Roman soldier slew the great man when Archimedes refused to hurry with him to Marcellus, because of his absorption in a mathematical problem. Archimedes's genius, courage, and love of mathematics will be celebrated as long as civilization lasts.

Although Alexandria remained a cultural center until its library was burned in 641 A.D., few advances were made in Greek mathematics after the days of Apollonius and Archimedes.[19]

No longer pursued in medieval Europe, mathematics continued to be cultivated by the Chinese, Indian, and Arabic peoples. In the 12th century, Arabic and Greek versions of the original Greek texts began to appear in Latin translation in Europe. Mathematics, philosophy and the study of nature all started to blossom once more. But, in a sense, the rigor of the Greek mathematics had to be given up before progress could be made. The numerical and algebraic side of mathematics had to be expanded despite obstacles that must have seemed insurmountable to the meticulous.

Reunification of Geometry and Algebra

There are many ways of specifying geometrical points. For example, we can identify the point A in Fig. 9 by specifying that it is the point of intersection of two given straight lines, KL and MN. Likewise, points B and C can be identified as the points of intersection of a given straight line and a given circle, while E, F, G can be identified as points of intersection of three given straight lines. This is a geometrical mode of identifying points.

Another way of specifying points is to use numbers as labels for them. This ingenious device is one of the most fruitful ideas in mathematics, and is indispensable for engineering and physics. The ancient Chinese and Babylonian mathematicians freely associated numbers with points. But, the discovery that the diagonal and side of a unit square are incommensurable led the meticulous Greek geometers to banish number from geometry. In hindsight, their understanding of number was much too restrictive and this severely limited the development of their mathematics.

A full comprehension of the real number system did not emerge until the 19th century, but the value of associating points with numbers was recognized in the 17th century by Fermat[20] and Descartes,[21] and was used creatively by these authors as a means of employing algebra to study geometry. In effect, Descartes and Fermat put number back into geometry. They reunited the algebraic and geometric traditions. Of

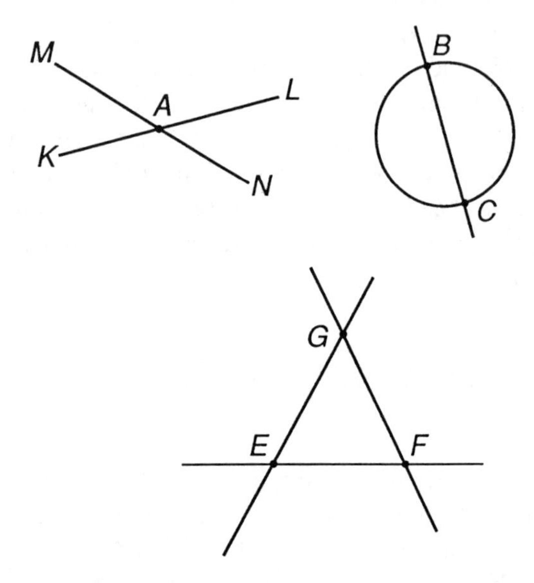

Figure 9 Geometrical specifications

course, they overlooked the logical difficulties which had been taken so seriously by the Greeks. Eventually, mathematicians were obliged to face these difficulties, and the resulting clarifications in the 19th century led to an unprecedented level of understanding of real numbers. After this, one no longer hesitates to assign numbers to points. Let us review how the numerical scheme of specification may be carried out.

Cartesian Coordinates

Confining attention at first to the Euclidean plane, pick an arbitrary point O (called the *origin*), draw an arbitrary straight line PQ through O, and then another straight line RS through O and perpendicular to PQ (Fig. 10). Choosing a line segment OI of any (nonzero) length, we assign the number 1 to OI. This is called fixing the unit of length. Take a line segment OJ also of length 1 unit. To each point on PQ to the right of O, we assign the number which represents the distance of that point from O in the chosen unit. For points on PQ, but to the left on O, we use distances preceded by a minus sign. The number 0 is assigned to O itself. Thus, to each point on PQ, a unique real number is assigned. Furthermore, for any given number, there is one and only one point on PQ which has been assigned that number. We have now established what is called a one-to-one correspondence between the points on PQ and the real numbers. Similarly, there is a one-to-one correspondence between the points on RS and the real numbers.

Now, take any point A in the plane and drop perpendiculars AM

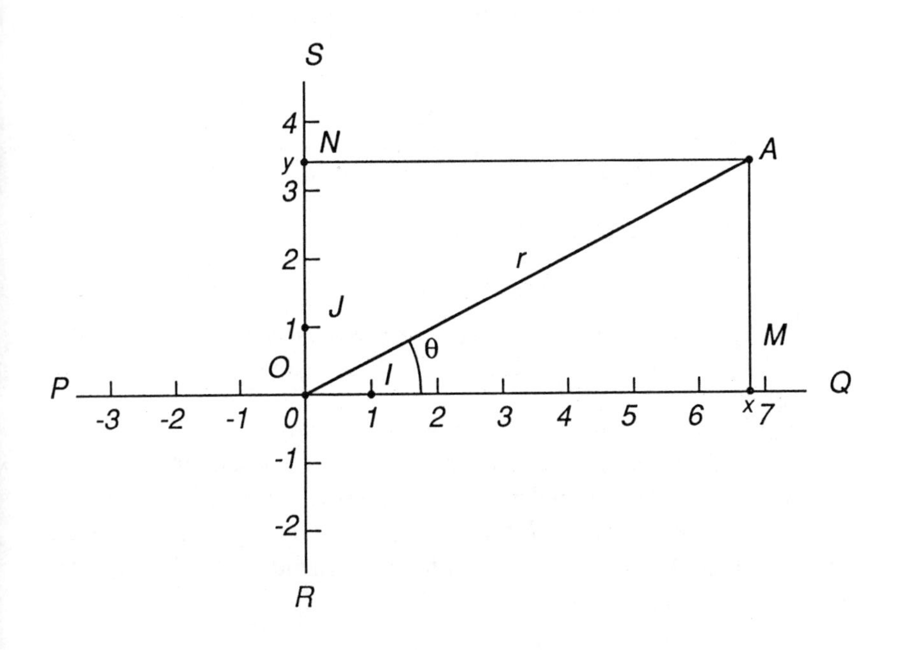

Figure 10 The system of rectangular Cartesian coordinates

and AN from A to the lines PQ and RS, respectively. Let x and y be the numbers previously assigned to the points M and N, respectively. These two numbers are the *rectangular Cartesian coordinates* of A—rectangular because RS is perpendicular to PQ, and Cartesian in honor of Descartes.[22] We often refer to PQ as the "x-axis" and RS as the "y-axis", and we also use the terminology "horizontal" and "vertical" axes, respectively, for them.

Using the foregoing procedure, we may assign a unique pair (x, y) of numbers, taken in that order, to any point A on the plane. Conversely, to each pair (x, y), there corresponds a unique point on the plane. We now have a one-to-one correspondence between points on the plane and number-pairs.

The distance r from the origin to A is equal to the length of the straight line OA, which according to Pythagoras's theorem is

$$r = \sqrt{x^2 + y^2} \, . \tag{1.18}$$

We also note the trigonometric relations

$$x = r \cos \theta \, , \quad y = r \sin \theta \, , \tag{1.19}$$

where θ is the angle that OA makes with the x-axis.

Having identified points by their coordinates, we may translate geometrical statements into algebraic ones and *vice versa*. For example, the geometrical statement "EF is a line passing through G and inclined at an angle θ (as shown in Fig. 11) to the horizontal direction" can be translated into the equation

$$y = mx + c \, , \tag{1.20}$$

where (x, y) are the coordinates of an arbitrary point A lying on EF, c is the ordinate of the point G, and

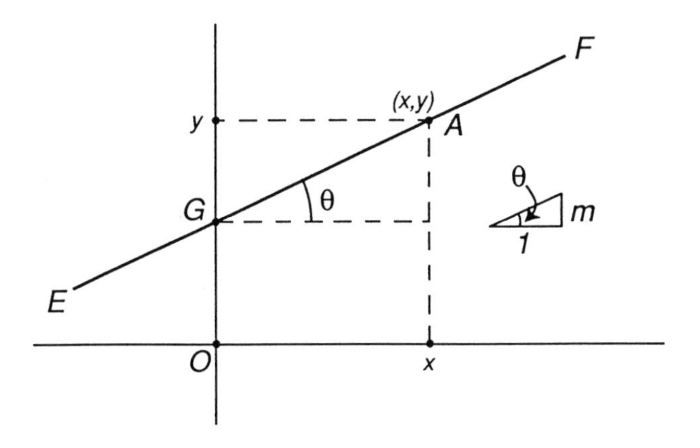

Figure 11 A straight line EF of slope m

$$m = \tan\theta \tag{1.21}$$

is the *slope* of EF.

Going in the reverse direction, we can easily see that any equation having the form of Equation (1.20), with given numerical values of m and c, corresponds to a unique straight line in the plane. This line is said to be the *locus* of points whose coordinates satisfy the given equation.[23]

The powerful tool, which Fermat and Descartes discovered, for converting geometrical statements into algebraic ones and *vice versa*, made tractable a vast complex of problems that were not amenable to pure geometrical methods, and opened up a whole new domain of questions concerning the loci produced by general algebraic equations. About this pivotal event in the history of mathematics, the British mathematician and philosopher, Alfred North Whitehead writes:[24]

> *... the essence of co-ordinate geometry is the identification of the algebraic correlation with the geometrical locus. ... We have thus arrived at a position where we can effect a complete interchange in ideas and results between the two sciences. Each science throws light upon the other, and itself gains*

immeasurably in power. It is impossible not to feel stirred at the thought of the emotions of men at certain historic moments of adventure and discovery–Columbus when he first saw the Western shore, Pizarro when he stared at the Pacific Ocean, Franklin when the electric spark came from the string of his kite, Galileo when he first turned his telescope towards the heavens. Such moments are also granted to students in the abstract regions of thought, and high among them must be placed the morning when Descartes lay in bed and invented the method of co-ordinate geometry.

In our discussion so far, x and y have signified rectangular Cartesian coordinates. By taking two coordinate axes through O that are not (or rather, not necessarily) at right angles to one another, one obtains *oblique Cartesian coordinates*. Instead of Fig. 10, we now have Fig. 12 in general. For convenience, we have taken the origin to be the same as before, and we have also used the same scale as before. We may regard the original axes PQ and RS as having been rotated independently about O into their new positions $\bar{P}\bar{Q}$ and $\bar{R}\bar{S}$, respectively. Let the angle between them be ϕ. The new coordinates of the point A are found as follows: the coordinate $\bar{x}$ is got by drawing a line through A parallel to the $\bar{y}$-axis and finding its intersection with the $\bar{x}$-axis; likewise, $\bar{y}$ is got by drawing a line through A parallel to the $\bar{x}$-axis and finding its intersection with the $\bar{y}$-axis. The coordinates $\bar{x}$ and $\bar{y}$ are the oblique Cartesian coordinates of A. Rectangular Cartesian coordinates are a special type of oblique Cartesian coordinates (corresponding to the choice $\phi = 90°$).

When oblique Cartesian coordinates are used, the expression for the distance from the origin to the point A does not, in general, have as simple a form as Equation (1.18). Instead, noting that the angle subtended by OA in Fig. 12 has measure $180-\phi$ degrees, and appealing to the cosine rule of trigonometry, we find that

$$r = \sqrt{\bar{x}^2 + \bar{y}^2 + 2\,\bar{x}\,\bar{y}\cos\phi}\ . \tag{1.22}$$

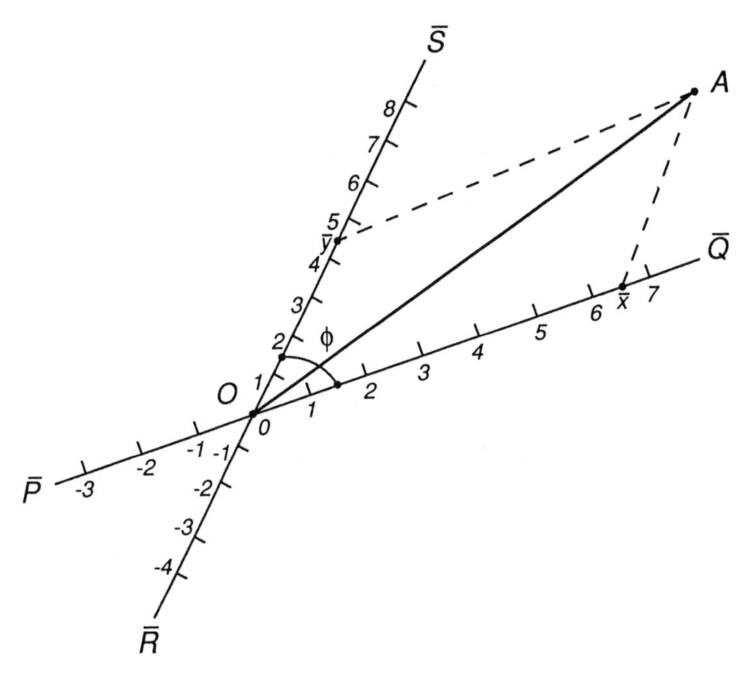

Figure 12 Oblique Cartesian coordinates

Let us see if we can come up with some other ways of assigning coordinates to points in the plane:

Experiment 2 (Coordinate systems): (**a**) On a sheet of paper, draw a large number of circles centered at a point O. Then, through O, draw a large number of straight lines. Pick a point P lying on the intersection of a circle of radius r (say) and a straight line inclined at an angle θ to the usual x-axis, and recall Equation (1.19). Can (r, θ) be used as coordinates?

(**b**) Could other figures be used instead of the circles in (**a**)?

(**c**) Could other curves be used instead of the straight lines in (**a**)? □

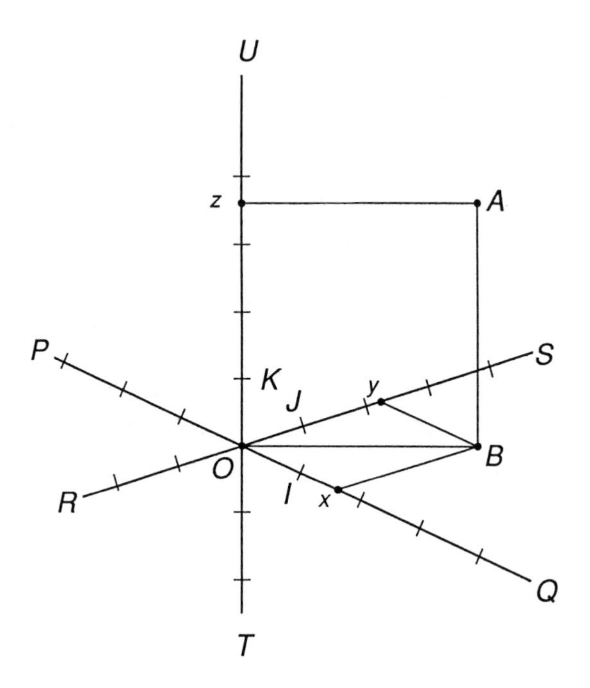

Figure 13 Cartesian coordinates in space

Next, we proceed to three-dimensional Euclidean space. There, we can set up a system of rectangular Cartesian coordinates as follows. Keeping the origin and scale to be the same as in Fig. 10, we may now regard the plane QOS to be a horizontal plane in space. At O, erect a straight line TU perpendicular to the plane QOS (Fig. 13). Choose a segment OK of unit length, and use it to assign numbers to the points on TU. Considering now any point A in space, we may drop a perpendicular from A to meet the plane QOS at some point B. Let z be the length of BA if A lies above QOS, and the negative of BA if A lies below QOS. We may call z the vertical coordinate of A. Next, dropping perpendiculars from B onto the lines PQ and RS, we obtain the x and y coordinates of B as in the planar case: these numbers also are the x and y coordinates of A. It should now be evident that we can assign a unique triple of numbers (x, y, z) to each point in space, and conversely that each triple generates a unique point; (x, y, z) are the rectangular Cartesian coordinates of A.

Just as in Experiment 2, more complicated coordinate systems can be set up in space as well. (Try using concentric spheres, planes, and cones.) General coordinate systems are extremely useful in mathematics and science, and we will have occasion to use such systems later.

Not long after the re-introduction of numbers into geometry by Descartes and Fermat, differential calculus was invented by Newton and Leibniz.[25] This major new tool extended the boundaries of mathematics in all directions. In geometry, it made possible the precise discussion of the local properties of general curves and surfaces, *i.e.,* the geometrical properties that they possess in the vicinity of a given point. This is the subject of differential geometry. You will be meeting many of the basic ideas in this fascinating field as the present book unfolds.

Notes to Chapter 1

[1] Our knowledge of the ancient history of mathematics has changed radically during the course of the 20th century. At the beginning of the century, it was believed that mathematics as a body of knowledge developed in Classical Greece. By the mid-1930s, due to the work of Otto Neugebauer, it became clear that a splendid mathematics existed in Old Babylonia in 1700 B.C. A description of this material can be found in Neugebauer (1957). In a long series of remarkable papers published during a 30-year period beginning in the late 1950s, the mathematician Abraham Seidenberg, of the University of California at Berkeley, argued on the basis of his historic and ethnographical studies that mathematics has a single origin in ancient ritual. He concluded that the content of ancient geometry is best represented by the Indian texts, the *Śulvasūtras* , which give detailed geometrical methods for the construction of altars. (See Seidenberg's papers of 1962, 1972, 1978, 1981, and 1983, referenced in the Bibliography.) B.L. van der Waerden, also a mathematician with a deep interest in the history of his subject, did not initially agree with Seidenberg's revolutionary views, but later embraced them and found further important evidence in support of Seidenberg's theory. The reader is referred to van der Waerden's excellent *Geometry and Algebra in Ancient Civilizations*, and also to Seidenberg's review of it

(Seidenberg, 1985) for an account of their views on the history of mathematics.

It is very difficult to assign a date to the Śulvasūtras . One estimate is that they were compiled sometime during the period 500-300 B.C. However, it must be borne in mind that these texts are only a codification of older knowledge. Seidenberg argues for the extreme antiquity of the latter. A translation of the *Śulvasūtras* may be found in Sen and Bag (1983).

[2] B.L. van der Waerden (1983), p. 29.

[3] See Heath (1921), Vol. 1, p. 130, as well as van der Waerden (1961), p. 87 and van der Waerden (1983), p. 87.

[4] In the opening pages of Vol. I of his magnificent *History of Greek Mathematics* (1921), T.L. Heath stresses the importance of foreign travel for the development of Greek philosophy.

[5] My account of Pythagorean teachings is based upon the following sources: Kirk and Raven (1957), Chapters VII and IX; Heath (1921), Vol. 1, Chapters III and V; I. Thomas (1951), Vol. 1, Chapters III and VI; *The Encyclopedia of Philosophy*, entry entitled "Pythagoras and the Pythagoreans"; *Dictionary of Scientific Biography*, entry entitled "Pythagoras of Samos"; Farrington (1953), Chapter III; *Dictionary of the History of Ideas*, entries entitled "Pythagorean Doctrines to 300 B.C." and "Pythagorean Harmony of the Universe."

[6] Proclus's summary of the development of Greek mathematics is translated in I. Thomas (1951), Vol. 1, Chapter 4. The excerpt is taken from p. 149.

[7] Aristotle, *Metaphysics* A5, 985 b 23 (the excerpt is based on the translations appearing in Heath (1921), Vol. I, pp. 66-67 and Kirk and Raven (1959), pp. 236-237).

[8] See Heath (1921), Vol. I, p. 143 and Thomas (1951), p. 177.

[9] Recall that two plane rectilinear figures are similar if and only if corresponding angles are equal and the sides about the equal angles are proportional. For two triangles, however, if corresponding angles are equal, then the triangles are

similar; further, if two triangles have their sides proportional, then the triangles are similar.

[10] See Heath (1921), Vol. 1, p. 91.

[11] For details, consult Chapters VI and VIII of Heath (1921), or refer to the entry on Zeno of Elea in the *Dictionary of Scientific Biography.*

[12] Heath (1921), Vol. 1, devotes a chapter to Plato. The articles on Plato in *The Encyclopedia of Philosophy* and *The Dictionary of Scientific Biography* are also excellent.

[13] The passage is from Book VI, 510 of *The Republic*, as translated by T. L. Heath on p. 290 of his *History of Greek Mathematics*, Vol. 1.

[14] A sequence of numbers $a_1, a_2, a_3, \ldots, a_n, \ldots$ possesses a *limit* a if and only if all members past some member a_M are arbitrarily close to a, *i.e*, for any given $\epsilon > 0$, there is a natural number M such that for $n > M$ the absolute value of $a - a_n$ is less than ϵ. The sequence is then said to *converge* to a or to *tend* to a. For example, the sequence $1, \frac{1}{2}, \frac{1}{3}, \ldots, \frac{1}{n}, \ldots$ tends to 0.

[15] For an English translation of Euclid's *Elements*, with copious notes and background information, see Heath (1926). Also, Heath (1921) has a long chapter on Euclid. In the *Dictionary of Scientific Biography*, there is a superb article by I. Bulmer-Thomas on Euclid, followed by an article on the fascinating story of the transmission of the *Elements* down through the ages.

[16] The statement of Definition 23 and Postulate 5 are from Heath's 1926 edition, Vol. 1, of Euclid's *Elements.*

[17] See Vol. II of Heath (1921) for details on Apollonius and Archimedes.

[18] This is one of Archimedes's assumptions in his *On the Sphere and Cylinder, Book 1*, a translation of which has been given by Heath (1897).

[19] There were political, cultural, and intrinsic mathematical reasons for this.

For an enlightening discussion, the reader is referred to Chapter VIII of van der Waerden (1961).

[20] Pierre de Fermat (1601-1665) was a French lawyer and mathematician. He did not write for publication, but discovered several important results, especially in number theory. His famous "last theorem", namely that there are no positive integers x, y, z such that $x^n + y^n = z^n$ for any positive integer n greater than 2, was recently proved by Andrew Wiles of Princeton University. (See *The New York Times,* 24 June 1993 and 31 January 1995.)

[21] René Descartes (1596-1650) was a brilliant French philosopher who made important contributions to mathematics and physics as well. His book on analytic geometry, *La Geometrie*, was published in 1637. (See Smith and Latham (1954).)

[22] Actually, the systems employed by Descartes and Fermat were not rectangular: they involved obliquely inclined lines.

[23] A *locus* is the set of all points that satisfy a given geometrical condition. For example, the condition "Q is any point of the plane lying a distance 1 from the origin O" has as its locus the unit circle centered at O. In space, the corresponding condition has a unit sphere as locus, while on a line the locus is just two points.

[24] This passage is quoted from Whitehead (1958), pp. 88-89.

[25] For an excellent history of the calculus, see Boyer (1949).

2

Basic Operations

> The world's most difficult undertakings
> necessarily originate while easy, and
> the world's greatest undertakings
> necessarily originate while small.
>
> . . .
>
> Contemplate a difficulty when it is easy
> Manage a great thing when it is small
>
> *Lao-Tzu*

We start out by performing some geometrical operations that can be easily done on objects located in ordinary three-dimensional space. You will need a ruler and a piece of string (or a tape measure) for measuring lengths, and a protractor for measuring angles.

Experiment 3 (Translations and rotations): Take a piece of stiff cardboard and on it draw a triangle, and a pair of parallel lines with a transversal. Measure the sides and angles of the triangle. Now slide the cardboard to a new location, and also rotate it. Check that the lengths and angles have not changed, and that the pair of lines are still parallel. □

This experiment illustrates that moving figures around by means of translations and rotations does not alter their basic measure relations. It can be shown theoretically that the only physical motions that preserve the distance between all points of a given figure in Euclidean space are combinations of translations and rotations. More specifically, every rigid motion can be effected by a translation followed by a rotation. (Alternatively, the rotation can be performed first.)

Experiment 4 (Bending): Take a sheet of paper and draw the same figures as in Experiment 3. Now tape the page onto a cylindrical container. Measure the sides and angles of the triangle, and calculate the angle sum. How far apart in space are the vertices of the triangle? □

You now see that even though the figures no longer lie in a plane, the original measure relations have been preserved; no change in the length of any side of the triangle has occurred, even though spatial distances have changed.

Experiment 5 (Stretching): Find a long, thin, and flat strip of elastic material on which figures can be drawn, and a smooth piece of board. (I used an elastic strip designed for exercising, and a small table.)

(**a**) Place the strip of material on the board and draw a rectangle together with its diagonals on it. Measure the lengths of the sides and the diagonals, and measure one of the angles included by the diagonals. Stretch the elastic material and secure it by passing the strip beneath the board and tying a knot at a convenient location. Measure the sides and diagonals of the new rectangle and measure the angle between the diagonals. Calculate the *stretch* (= new length/old length) of various lines. [I stretched a rectangle of approximate dimensions 9 cm by 6 cm into a rectangle measuring 24.7 cm by 3.7 cm. This gave a longitudinal stretch of 2.7 and a transverse stretch of 0.6 (*i.e.,* a contraction). Also, the angle subtending the shorter sides of the rectangle changed from about 68.5° to about 17°.]

(**b**) On the stretched strip, trace a circle and measure its radius. Now release the strip. Measure the dimensions of the figure which the circle has become. Try some other ideas of your own. □

The operations that were performed in Experiments 3,4,5 and all variations and combinations of them, are known collectively as *deformations*. They can be carried out physically without cutting, puncturing, overlapping, or gluing of material. We shall make much use of deformations throughout this book, and the reader is encouraged at this stage to come up with other examples.

Experiment 6 (Reflecting): Position yourself facing a wall mirror. Hold the back of your left hand against the wall to the left of the mirror, with fingers pointing upwards. Raise your right hand, palm facing the mirror, and move this hand back from the mirror until it is behind your field of vision. Compare your left hand and the image of your right hand.

How would you make a reflected version of a given physical object? If two objects are mirror images of one another, are distances between corresponding points of them the same? □

If two figures are related through a deformation, then, despite their not being congruent in general, they still share many properties. For example, points lying on the line segment joining two given points P and Q of the original figure will be deformed into points lying on a curve joining the new positions, or "images", of P and Q in the deformed figure. Furthermore, by taking points closer and closer to P, we are assured that their images lie closer and closer to the image of P. It is true, conversely, that by taking points closer and closer to the image of P in the deformed figure, the original positions of these points gets closer in the undeformed figure. These observations on closeness also hold – only more trivially – for rigid motions and reflections.

Experiment 7 (Cutting, joining): (**a**) Take two sheets of paper. On one, draw a triangle such as that indicated in Fig. 14. The point D is chosen approximately midway between A and B, and the segments P_iD are equal to DQ_i $(i = 0, 1, 2, ...)$, respectively. Draw the same figure on the other sheet. Cut the second sheet along the line CD and move the cut pieces apart from one another. Focussing now on the uncut figure, notice that as i increases, the distance $P_i\,Q_i$ approaches zero; the points P_i and Q_i approach one another, each moving towards D. What can you say about the sequence of images of these points on the cut pieces?

(**b**) The same setup can be used to study joining. Keep the two cut pieces apart physically, but imagine them to be joined in such a way as to re-create a triangle congruent to ABC on the uncut sheet. Identify some points that are far apart from one another in the cut figure, but which lie close to one another in the uncut figure. □

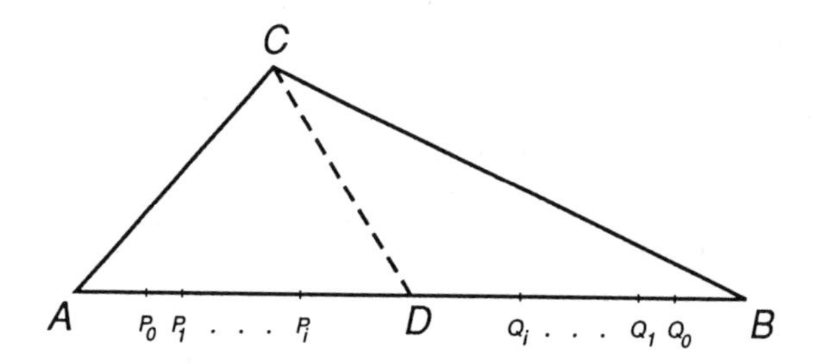

Figure 14 Make two triangles like this

Experiment 7 demonstrates that cutting can result in points losing their original closeness, while joining can bring non-close points close to one another. A combination of cuts and joins will, in general, only make the situation worse. However, there is a special cut-and-join operation that preserves closeness, as indicated in the next experiment.

Experiment 8 (A perfect cut-and-join): (**a**) Take two congruent, long strips of paper and mark them as shown in Fig. 15. Let us identify them as Strip No. 1 and Strip No. 2. Tape the ends of Strip No. 1 together to form a cylinder. For Strip No. 2, rotate the end CD one full turn $(360°)$ relative to the end AB, and then tape the ends together. In both strips, A coincides with D, and B coincides with C. We can regard Strip No. 2 as having been obtained from Strip No. 1 by cutting the cylinder along AB $(= CD)$, rotating one end, and joining in such a way that all points

Figure 15 Make two long strips and mark as shown

on the ends are brought back into contact with their original neighbors. Answer the following questions:

(1) Can the points on the two bands be made to correspond to one another such that: for each point on Strip No. 1, there is a unique point on Strip No. 2, and conversely, for each point on Strip No. 2, there is a unique point on Strip No. 1. More briefly, is there a one-to-one correspondence between points on the two strips?

(2) If two points (*e.g.*, P, Q) are close together on Strip No. 1, are they also necessarily close on Strip No. 2 (and conversely)?

(3) Can Strip No. 1 be deformed into Strip No. 2?

(**b**) Take two pieces of yarn, equal in length. Tie the ends of one piece together to form a circle. On the other piece, make a loose knot in the middle, and then tie its ends. Answer the same questions as in **(a)**. □

We now introduce a fundamental concept, namely that of two figures being *topologically equivalent* to one another, and will consider it in more detail in Chapter 5. Two essential properties are required for this equivalence to hold, and we have encountered both of them already: (a) there must be a one-to-one correspondence between points of the two figures; (b) the closeness property must hold in both directions; that is to say, closeness of points in one figure must imply closeness in the other figure, and vice versa. If two figures can be obtained from one another by a deformation, then they are automatically topologically equivalent. Thus, topologically speaking, a soccer ball, a rugby ball, and an American football are all equivalent. Similarly, circles, squares, and hexagons are topologically equivalent to one another.

Experiment 8 demonstrates that two figures can be topologically equivalent without being obtainable from one another by a deformation: the two bands are topologically equivalent to one another, but neither can be deformed into the other (no cutting is allowed in a deformation); likewise, the yarn circle is topologically equivalent to the knotted loop, regarded as a curve in space. Another example is shown in Fig. 16; on

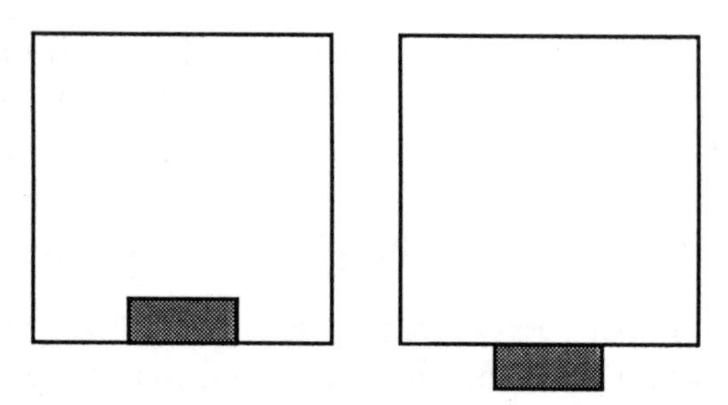

Figure 16 The two figures are topologically equivalent, but cannot be deformed into one another

the left, we have a block inside a closed box, and on the right is the box placed on top of the block. The one-to-one correspondence property and the closeness property are both satisfied by the two figures. However, the figures cannot be deformed into one another. (Opening the box would require cutting or tearing–which are not allowed in a deformation.)

Thus, in summary, while every deformation yields topologically equivalent figures, not every pair of topologically equivalent figures is obtainable by a deformation: perfect cut-and-join operations are also allowed. By itself, cutting is not a topological operation. Neither is joining. What distinguishes the perfect cut-and-join operation exemplified in Experiment 8 is that the original local geometrical environment of each point is restored, even though the global geometry of the figure has been changed.

3

Intersecting with a Closed Ball

In order to examine the local geometry of a figure, we will need to focus attention on the collection of points that lie close to any given point. We will dissect the figure mentally, as it were. To be clear about what this means, it is best to employ some concepts from set theory.

First some basic ideas, notation, and terminology: If an object P belongs to a set $\mathcal{S}$ (for example, if a point lies on a geometrical figure), we write $P \in \mathcal{S}$. If every member (or element) of a set $\mathcal{T}$ also belongs to $\mathcal{S}$, we write $\mathcal{T} \subseteq \mathcal{S}$ and say $\mathcal{T}$ is a *subset* of $\mathcal{S}$; if $\mathcal{T} \subseteq \mathcal{S}$, but $\mathcal{S}$ has at least one element not belonging to $\mathcal{T}$, we write $\mathcal{T} \subset \mathcal{S}$ and call $\mathcal{T}$ a proper subset of $\mathcal{S}$. If $\mathcal{T} \subseteq \mathcal{S}$ and also $\mathcal{S} \subseteq \mathcal{T}$, then $\mathcal{S} = \mathcal{T}$: they have the same members.

The *intersection* of any two sets $\mathcal{S}$ and $\mathcal{T}$ is the set consisting of all those elements that belong to both $\mathcal{S}$ and $\mathcal{T}$; it is denoted by $\mathcal{S} \cap \mathcal{T}$. The *union* of $\mathcal{S}$ and $\mathcal{T}$ is the set consisting of all the members of $\mathcal{S}$ plus all those members of $\mathcal{T}$ that do not already belong to $\mathcal{S}$; it is denoted by $\mathcal{S} \cup \mathcal{T}$. If $\mathcal{S}$ and $\mathcal{T}$ have no members in common, they are said to be *disjoint*, and we write $\mathcal{S} \cap \mathcal{T} = \varnothing$ and call $\varnothing$ the *empty set.*

Example 1: Consider the following three subsets of real numbers:

$$\begin{aligned} \mathcal{A} &= \text{set of numbers greater than 0 but less than 2} \\ &= \{x : 0 < x < 2\}, \\ \mathcal{B} &= \{x : 1 < x \leq 3\}, \\ \mathcal{C} &= \{1\}. \end{aligned}$$

We then have

$$
\begin{aligned}
\mathcal{A} \cap \mathcal{B} &= \{x : 1 < x < 2\}, \\
\mathcal{A} \cup \mathcal{B} &= \{x : 0 \leq x \leq 3\}, \\
\mathcal{A} \cap \mathcal{C} &= \mathcal{C}, \\
\mathcal{A} \cup \mathcal{C} &= \mathcal{A}, \\
\mathcal{B} \cap \mathcal{C} &= \varnothing, \\
\mathcal{B} \cup \mathcal{C} &= \{x : 1 \leq x \leq 3\}.
\end{aligned}
$$

You can also form the sets $\mathcal{A} \cap (\mathcal{B} \cap \mathcal{C})$, $\mathcal{A} \cap (\mathcal{B} \cup \mathcal{C})$, *etc.*

We will make frequent use of certain subsets of a line, a plane, and 3-dimensional space. On a line, the set of all points lying between two distinct points A, B and including A and B is called the *closed interval* with end-points A and B; if A and B are excluded from the set, the interval is called *open.* In a plane, the set of all points contained in a circular region of radius $r > 0$, centered at a point P, is called a *closed disk*; if the points lying on the circumference are excluded, the disk is said to be *open.* In space, the set of all points contained in a spherical region of radius $r > 0$, centered at a point P, is called a *closed ball*; if points on the sphere are excluded, the ball is said to be *open.* For an open ball of radius r, centered at P, we use the notation $\mathcal{B}_r(P)$, and to indicate a closed ball, we write $\overline{\mathcal{B}}_r(P)$.

A useful method of examining the local geometrical properties of an object is to intersect it with closed balls of arbitrarily small size centered at various points of interest. Consider, for example, the solid half-cylinder in Fig. 17. If we take a point E deep in the interior of the solid and intersect the solid with a closed ball of small radius and center E, we obtain a closed ball of points of the solid. At a point F lying on the top face of the half-cylinder, intersecting with a closed ball would yield a half-ball of points of the solid (points belonging to the surface of

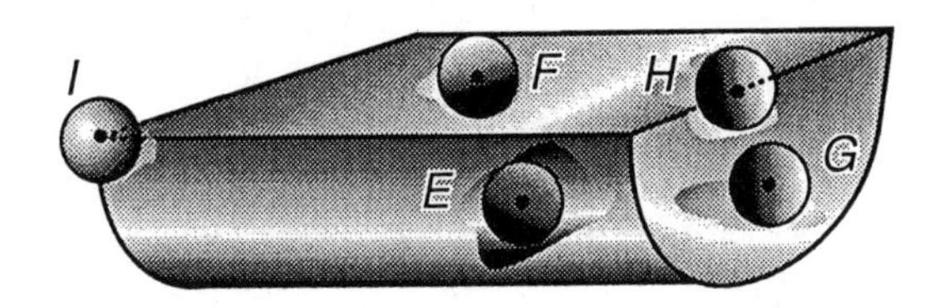

Figure 17 Using closed balls to examine local geometry

the hemisphere being included). Intersecting with a closed ball at a point G on the right-hand flat side of the solid would also yield a half-ball. Intersecting at H would produce a quarter-ball. At a vertex such as I, an even smaller portion of a ball would result from intersection with a closed ball.

Experiment 9 (Closed disks): On a sheet of paper or cardboard, draw some figures. Choose various points on the figures and construct a small circle at each, representing the boundary of a closed disk. Color the intersections made by the figure and the closed disks. Compare these intersections to one another. □

We may also use closed balls to systematically take a figure apart. Consider, for example, a straight-line segment 5 cm long. Let closed balls of radius 0.5 cm be constructed at points located 0.5, 1.5, 2.5, 3.5, and 4.5 cm from one end. The intersections of these balls and the original segment (considered as a closed interval) are the five closed intervals [0,1] ($= \{x : 0 \leq x \leq 1\}$), [0,2], [2,3], [3,4], [4,5]. We may call this a *dissection* of the figure. It is clear that the original segment can be reconstructed by taking the union of the five unit segments in the proper order. A less neat, but nevertheless serviceable, dissection of the line-segment can be gotten by placing balls of radius 1 cm at points located 1, 2, 3, 4 cm from an end. The resulting four closed intervals, namely [0,2], [1,3], [2,4], [3,5] overlap one another sequentially, but their union still reproduces the original line-segment.

Experiment 10 (Dissections): Draw an arbitrary figure on a sheet of paper and construct a family of disks that dissect it. Can you also dissect your figure using other elementary shapes, such as triangular or rectangular "disks"? Can you use disks of arbitrary shape, possibly not even all the same? How does this compare to cutting a figure into pieces of arbitrary shape? □

4

Mappings

We now discuss the concept of a mapping (or function). The usefulness of this idea for the mathematical sciences can hardly be exaggerated.

Let $\mathcal{A}$ and $\mathcal{B}$ be two nonempty sets, and suppose that we have a means of associating elements of $\mathcal{B}$ with elements of $\mathcal{A}$ such that the following condition is met: to each element a of $\mathcal{A}$, there is associated one and only one element b of $\mathcal{B}$. We then have a *mapping* (or *function*) from $\mathcal{A}$ into $\mathcal{B}$. The symbol f is used to denote a generic function, and we write $f : \mathcal{A} \to \mathcal{B}$, which may be read as "$f$ maps $\mathcal{A}$ into $\mathcal{B}$". Further, b is called the *image* of a under the mapping f (or the *value* of the function f at a) and the notation $b = f(a)$ [read: "f of a"] is used universally to signify this. The set $\mathcal{A}$ is called the *domain* of the mapping, while $\mathcal{B}$ is called the *co-domain*. Note that it is legitimate to have one element of $\mathcal{B}$ assigned to two different elements of $\mathcal{A}$, and also to have elements in

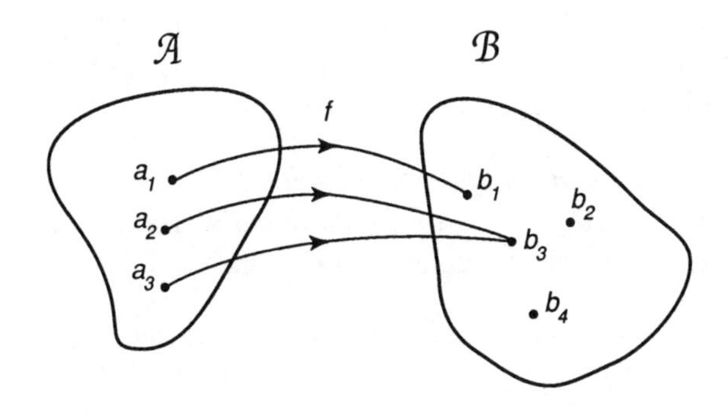

Figure 18 A mapping f from a set $\mathcal{A}$ into a set $\mathcal{B}$

$\mathcal{B}$ that are not associated with any element of $\mathcal{A}$. All that is required is that every element of $\mathcal{A}$ be assigned a unique element of $\mathcal{B}$. The general situation is represented pictorially in Fig. 18: a_1 is assigned a unique element b_1; a_2 is assigned a unique element b_3; a_3 is assigned a unique element b_3. If the set $\mathcal{B}$ is equal to $\mathcal{A}$, we just call f a mapping *on* $\mathcal{A}$.

If $\mathcal{W}$ is a subset of the domain $\mathcal{A}$, we use the notation $f(\mathcal{W})$ to signify that subset of the co-domain $\mathcal{B}$ which consists of the images of the elements of $\mathcal{W}$. The set $f(\mathcal{A})$ is called the *range* of f.

We will give several examples to illustrate the concept of a function.

Example 1: The function w that tells the birth weight of each person in a given population. Here, the domain is the given set of persons, and as co-domain we may take the set of positive real numbers (or any set containing them). Since each person has a unique birth weight, a mapping is defined. The range of w is the set of birth weights recorded.

Example 2: The mapping $f(x) = x^2$, which gives the square of each real number. The domain of this function is the set of real numbers; as co-domain we may take the set of nonnegative real numbers, which is actually the range of the function.

Example 3: The speedometer function, which gives a reading of a vehicle's speed at each instant during a trip. The domain can be chosen as the closed interval $[0, T]$ of real numbers corresponding to the interval of time during which the trip occurs. As co-domain, we may take the set of nonnegative real numbers. The (legal) range is a subset of this, equal to the closed interval from 0 to the speed limit.

Example 4: The constant mapping, which takes every element of the domain into a single element of the co-domain. The set consisting of this single element is the range of the function.

Example 5: The identity mapping id on $\mathcal{A}$, $a = id(a)$. *id* maps each element into itself.

Example 6: The mapping (let's call it *cart*) that assigns rectangular Cartesian coordinates to each point in space:

$$(x, y, z) = cart(P) \ . \tag{4.1}$$

The domain of *cart* is the set of points in space, and its range is the set of triples of real numbers. Equivalently, we may define three *coordinate functions* $\hat{x}$, $\hat{y}$, and $\hat{z}$ such that

$$x = \hat{x}(P) \ , \quad y = \hat{y}(P) \ , \quad z = \hat{z}(P) \ . \tag{4.2}$$

The device of using a hat (or some similar decoration) over a symbol to distinguish a function from its value is a convenient one, and we will employ it frequently.

A mapping $f : \mathcal{A} \to \mathcal{B}$ is called *one-to-one* (or *one-one*, or *injective*) if no element of $\mathcal{B}$ is assigned to more than one element of $\mathcal{A}$ (*i.e.,* if $f(a_1) = f(a_2)$ implies that $a_1 = a_2$); it is called *onto* if every element of $\mathcal{B}$ is assigned to at least one element of $\mathcal{A}$. The mapping in Fig. 18 is neither *one-to-one* nor onto. If a mapping is both one-to-one and onto, it is said to form a *one-to-one correspondence* between the elements of $\mathcal{A}$ and $\mathcal{B}$. Such a mapping is also called *bijective*. In this case, to each element of $\mathcal{A}$ is associated a unique element of $\mathcal{B}$, and to each element of $\mathcal{B}$ is associated a unique element of $\mathcal{A}$. Consequently, when f is bijective, there always exists a mapping from $\mathcal{B}$ to $\mathcal{A}$ that assigns to $b \in \mathcal{B}$ the element of $\mathcal{A}$ that is mapped into b by f; this function is called the inverse of f and is denoted by the symbol f^{-1}.

An important application of bijective functions occurs in one of the most basic processes of mathematics, namely in counting:

Example 7: Let $\mathcal{N}_k (k \geq 1)$ denote the subset $\{1, 2, 3 \ldots, k\}$ of the natural numbers $\mathcal{N}$. A set $\mathcal{A}$ is said to be *finite* if and only if it can be put into one-to-one correspondence with one of the subsets $\mathcal{N}_k$, or if it is empty. If $\mathcal{A}$ is not finite, it is called *infinite*. An infinite set $\mathcal{A}$ is *denumerable* if and only if it can be put into one-to-one correspondence with the set $\mathcal{N}$. For instance, the even natural numbers are denumerable, because the function $f(n) = 2n$ is a bijective function from $\mathcal{N}$ into the set of even positive integers. The set of integers is denumerable: they can be arranged as $0, 1, -1, 2, -2$, *etc*. The set of rational numbers is also denumerable. The set of irrational numbers is infinite but not denumerable, and so is the set of real numbers.[1] In fact, the set of real numbers in the closed interval $0 \leq x \leq 1$ is not denumerable either. Furthermore, it can be shown that each of the intervals $0 < x < 1$, $0 \leq x < 1$, and $0 < x \leq 1$ can be put into one-to-one correspondence with the closed interval $0 \leq x \leq 1$.

From Example 7, it is evident that the set of real numbers is vastly richer than the set of rationals. We take this opportunity to mention a subtle but essential property that the real number system possesses (but the rational number system does not). Consider, for instance, the set $\mathcal{A}$ of nonnegative real numbers less than the (irrational) number $\sqrt{2}$: $\mathcal{A} = \{x : 0 \leq x < \sqrt{2}\}$. This set is bounded above and below, *i.e.,* there is a real number b such that every element of $\mathcal{A}$ is less than or equal to b (*e.g.,* $b = 2$), and likewise there is a real number c such that every element of $\mathcal{A}$ is greater than or equal to c (*e.g.,* take c equal to -1). The real number system has the following property: if a nonempty subset of real numbers is bounded above, then one of its upper bounds is less than all the others; it is called the *least upper bound* (or *supremum*). Likewise, every nonempty subset of real numbers that is bounded below possesses a *greatest lower bound* (or *infimum*). For the set $\mathcal{A}$ just mentioned, the least upper bound is $\sqrt{2}$, and the greatest lower bound is 0. If the least upper bound of a set is actually a member of the set, then it is the maximum number in the set; likewise, if the greatest lower bound is a member of a set, then it is the minimum number in the set. For the set $\mathcal{A}$, the least upper bound does not belong to $\mathcal{A}$, but the greatest lower bound does. Note that for the set of rational numbers that are greater than or equal to zero and less than the (irrational) number $\sqrt{2}$, there is no *rational* upper bound that is less than all other upper bounds: the set of rationals does not satisfy the supremum property.

We turn now to two experiments in which you will generate functions from measurements.

Experiment 11 ("Falling ladder"): Let us represent a ladder leaning against a vertical wall by a line of length 15 cm drawn so that one end A touches the vertical (y) axis on a sheet of squared-paper and the other end B touches the horizontal (x) axis.

(**a**) For several measured inclinations (θ) of the line, measure the horizontal and vertical intercepts, and tabulate. Use the trigonometric functions on your calculator to determine the level of accuracy that can be expected in measuring the functions $x(\theta)$ and $y(\theta)$ experimentally.

(**b**) For every inclination of the line, erect a vertical where the line intersects the horizontal axis, and erect a horizontal where the line intersects the vertical axis, and find the point of intersection of these perpendiculars. What is the locus of these points ? □

Experiment 12 (Stretch function): Take an elastic strip like the one you used in Experiment 5. From a point on the strip draw radii of length 8 cm (say) at several different angles (θ). Now stretch the strip and secure it. Measure the new lengths of the radii and plot the stretch λ as a function of θ. □

Besides the basic definition of a function that was given above, there are other useful ways of thinking about this concept. Some of these are brought out in following examples.

Example 8: The straight line. Recalling Equation (1.20), we note that the locus EF in Fig. 11 is produced for fixed values of m and c by letting x take on all possible values and by employing Equation (1.20) to calculate the corresponding values of y. Note that for each value of x, there is one and only one value of y. A mapping is therefore defined. We say that Equation (1.20) prescribes y as a function of x, and we call x the *independent variable* and y the *dependent variable.*[2]

It is often profitable to regard Equation (1.20), and similar equations more general than it, from the following kinematical point of view: Let x and y represent the coordinates of a material point (an idealized marble) at time t, and suppose that the material point moves in such a way that x and t are numerically equal, *i.e.*,[3]

$$x = t\,. \tag{4.3}$$

Then, by virtue of Equation (1.20), the value of y at time t must be

$$y = mt + c\,. \tag{4.4}$$

Thus, as time progresses, the particle moves out along the line EF, occupying the point G at $t = 0$ and arriving at the point A at time t. In other words, the locus can be regarded as being generated by the material point, moving in accordance with Equations (4.3) and (4.4). This is a very suggestive viewpoint.

When we replace an equation like Equation (1.20) by a pair of equations like Equations (4.3) and (4.4), we say that the locus has been *parametrized* by the new variable, or *parameter*, t. Time is often a convenient parameter, but as we shall see, others are useful too.

Example 9: The circle. Consider a circle of radius $R(> 0)$ centered at the origin (Fig. 19). We can think of this circle in a static way as the locus of points lying a distance R from the origin. Alternatively, we can think of it in kinematical terms as the curve traced out by a material point moving in such a way that its distance from O is always R. Consider an arbitrary point A on the circle and let (x, y) be its coordinates. Join OA and let θ be the angle (measured in radians) that OA makes with the positive half of the x-axis. The angle θ can be any real number that is greater than or equal to zero and less than 2π $(0 \leq \theta < 2\pi)$. The coordinates x and y can be written in terms of R and θ as

$$x = R\cos\theta\,, \quad y = R\sin\theta\,. \tag{4.5}$$

Of course,

$$x^2 + y^2 = R^2\,. \tag{4.6}$$

If we solve Equation (4.6) for y, we see that

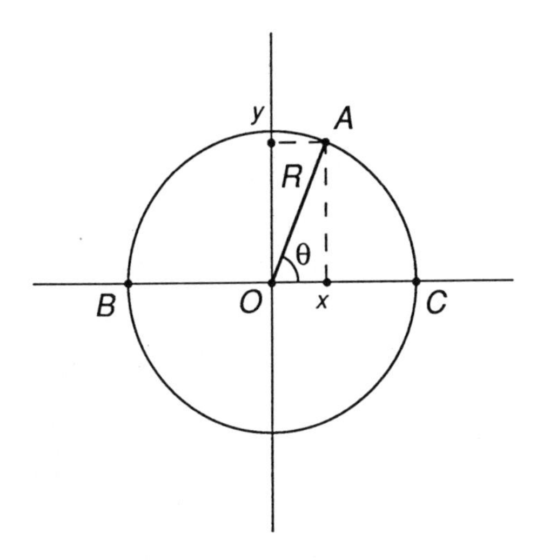

Figure 19 The circle

$$y = \pm\sqrt{R^2 - x^2} \ . \tag{4.7}$$

The latter equation does not represent a function, because there exist values of x which when substituted into Equation (4.7) yield two, and not one, value of y. Even if there was only one value of x that did this, it would be enough to cause Equation (4.7) not to represent a function; but, in fact, all values of x lying in the open interval $-R < x < R$ produce two values of y. These correspond to the two points at which a vertical line drawn through the point whose coordinates are $(x, 0)$ intersects the circle.

We can salvage the situation somewhat by considering not the circle as a whole, but instead two disjoint portions composing it, namely the piece which lies above the x-axis but also includes the two points B and C, and the piece that lies below the x-axis (Fig. 20). The curve in Fig. 20a is described by the Equations (4.5) but with θ restricted to lie in the interval $0 \leq \theta \leq \pi$, which makes $y \geq 0$. Similarly, the curve in Fig. 20b is described by Equations (4.5) but with $\pi < \theta < 2\pi$, which

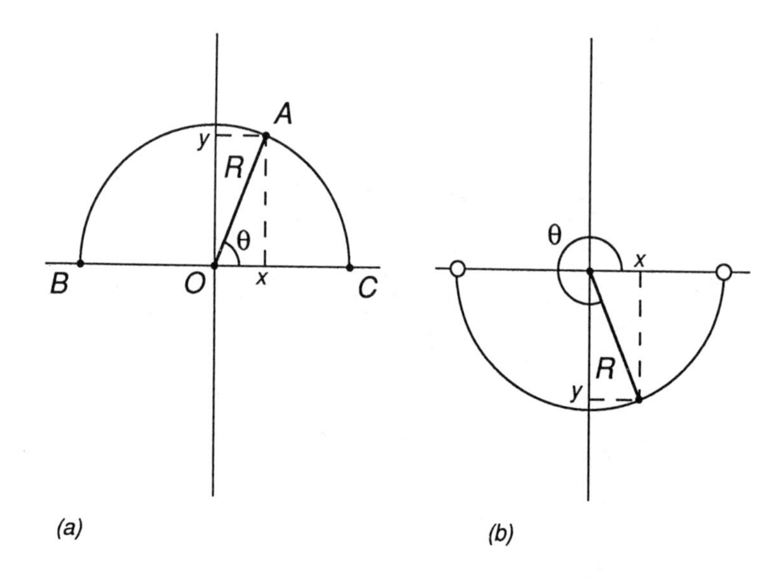

Figure 20 Decomposing a circle into two disjoint pieces

makes $y < 0$. Equation (4.6) holds for both curves, but with $y \geq 0$ for the curve in Fig. 20a and $y < 0$ for the one in Fig. 20b. Consequently, when we solve for y, for the curve in Fig. 20a we get

$$y = + \sqrt{R^2 - x^2}\,, \tag{4.8}$$

whereas for the curve in Fig. 20b, we get

$$y = - \sqrt{R^2 - x^2}\,. \tag{4.9}$$

Each of these two equations describes a function, which is what we have been seeking. Thus, the circle is the union of two disjoint pieces each of which can be represented by an equation that yields y as some function of x.

Figure 21 Another curve for which y is not a function of x

The situation encountered in Example 9 is typical: for general planar curves, a single function taking x-values into y-values would not suffice. The curve in Fig. 21 another example. In such cases, the kinematical and parametrical descriptions become very useful. Returning to the circle, let us take θ numerically equal to time ($\theta = t$). Then, starting our stopwatch from zero and letting it run, we see from Equations (4.5) that a material point occupies the point C at $t = 0$, and proceeds steadily in a counterclockwise direction, arriving at B when $t = \pi$, and returning to C at $t = 2\pi$. To occupy all points of the circle and each point only once, we limit out parameter t to the interval $0 \leq t < 2\pi$.

Example 10: The ellipse. Several alternative definitions can be given for the ellipse. For our purposes, the following one is the most convenient (Fig. 22): Set up a rectangular Cartesian coordinate system in the plane. With center O, draw two circles of radii a and $b < a$. Take any line through O, inclined at an angle θ to the positive x-axis and intersecting the circles in M and N, respectively. Through N draw a line parallel to the x-axis, and through M draw a line parallel to the y-axis: let P be the point of intersection of these lines, and let (x, y) be its coordinates. Denote the point at which the vertical MP meets the x-axis by S. Also, through N draw a vertical, meeting the x-axis at some point T.

By construction,

$$OM = a \; , \quad ON = b \; . \tag{4.10}$$

As can be seen with the help of Fig. 22, the x-coordinate of P is given by

$$x = a \cos \theta \; . \tag{4.11}$$

Similarly, the y-coordinate of P is

$$y = b \sin \theta \; . \tag{4.12}$$

Hence, recalling the trigonometric identity $\sin^2\theta + \cos^2\theta = 1$, we conclude from (4.11) and (4.12) that

$$\frac{x^2}{a^2} + \frac{y^2}{b^2} = 1 \; . \tag{4.13}$$

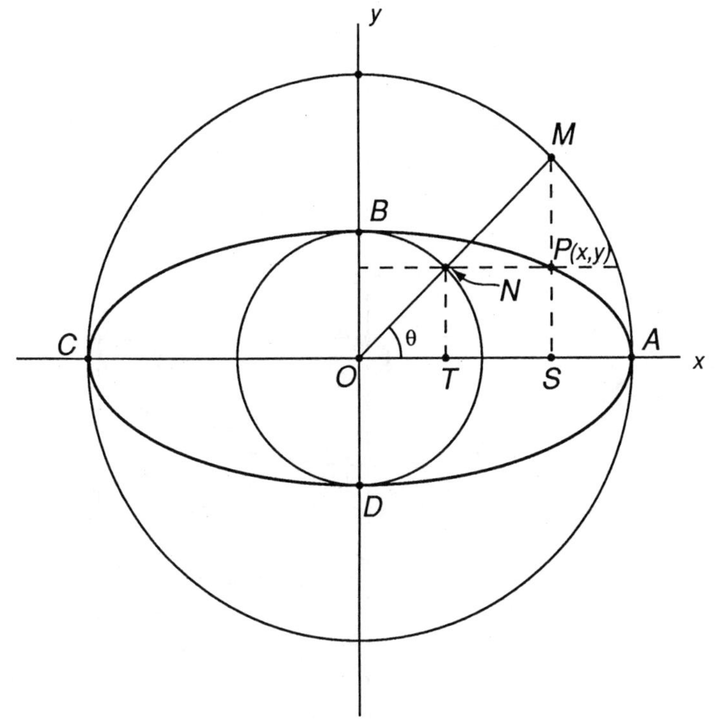

Figure 22 The ellipse

Equation (4.13) is one form of the equation of an ellipse. Observe that the points $A[x = a, y = 0], B, C, D$ lie on the ellipse. The line segment CA is called the *major axis* of the ellipse, and has length $2a$; likewise, BD is the *minor axis* and has length $2b$. Note that if a had been taken equal to b in our construction, a circle would have resulted, and that if a is set equal to b in Equation (4.11), an expression of the form (4.6) is obtained.

An important property of the ellipse is explored in the next experiment.

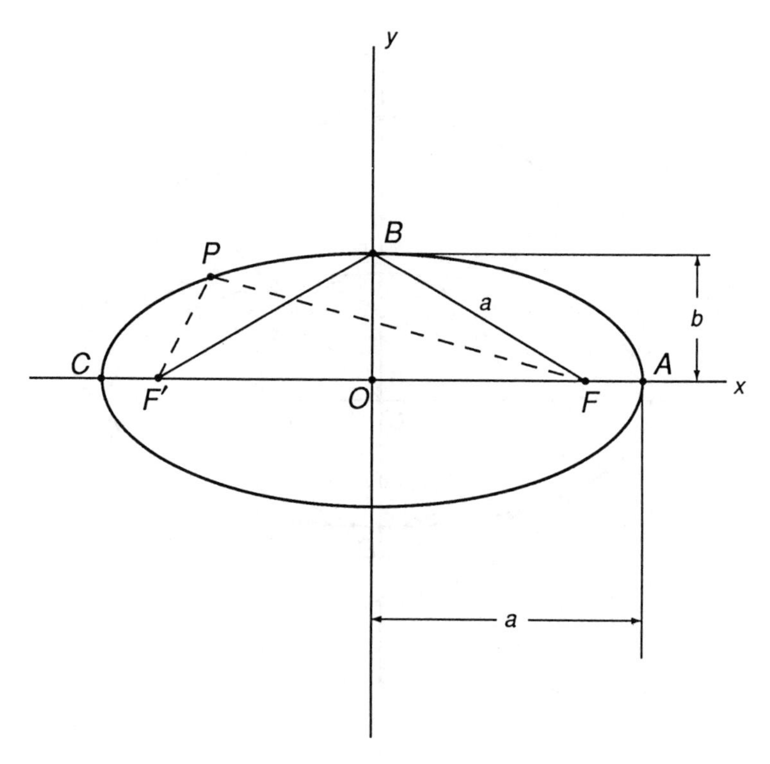

Figure 23 A property of the ellipse

Experiment 13 (Ellipse): (**a**) Choose lengths a and b and determine several points on the ellipse $ABCD$ by means of the construction described in Example 10. Now, with center B and radius a, draw a circle. This will intersect the x-axis at points F' and F, as indicated in Fig. 23. Each of these is called a *focus* of the ellipse. Note that $OF = OF' = \sqrt{a^2 - b^2}$ ($= c$, say).

(**b**) For each point P that you found on the ellipse, join P to the two foci, and measure the segments PF and PF'. Tabulate the sums $PF + PF'$. What do you conclude?

(**c**) The property alluded to in (**b**) suggests the following method of drawing an ellipse with semimajor axis of length a and semiminor axis of length b: Choose two points F, F' a distance $2c$ apart on a sheet of paper. Secure the paper to a wooden board and hammer in a small nail at F and at F'. Make a loop of thread of length $2(a + c)$. (A good way of making the loop is to hammer two small nails in the wood a distance $a + c$ apart, loop the thread tautly around them, tie a knot, trim the ends, and remove the nails.) Place the loop over the nails at F, F' and pull the loop taut with the tip of a pencil. Move the pencil tip about on the paper while keeping the loop taut. (If your pencil keeps slipping out of the loop, use two strands of thread.) □

Notes to Chapter 4

[1] Clarification of these concepts did not come about until late in the 19th century, when G. Cantor (1845-1918) developed his revolutionary and brilliant theory of infinite sets.

[2] In the present case, since Equation (1.20) can be solved (provided $m \neq 0$) to yield

$$x = \frac{1}{m} y - c ,$$

it is also true that x is a function of y (but not the same function as y is of x). In the new function, y is the independent variable and x is the dependent one.

[3] The value $t = 0$ can be chosen to correspond to some convenient instant of time. For example, we might set a stopwatch to zero at 12 noon. Negative values of t then correspond to a.m. times, and positive values to p.m. times.

5

Preserving Closeness: Continuous Mappings

In discussing the operations described in Chapter 2, we paid particular attention to whether or not closeness between points of a figure was changed by the operations. It is worth looking into this idea more carefully.

In a general set, there is no natural notion of distance between elements. (Think, for example, about the set whose elements are the word cat, the color red, and the number π.) But, in the case of geometric sets, with which we are primarily concerned in this book, we can always speak sensibly about elements of a set being close to one another or not. Before we concentrate on the geometrical case, let us do a little numerical experiment:

Experiment 14 (Trying to get closer): (**a**) Consider the function $y = f(x) = x^2$, and sketch its graph. Choose an arbitrary value of x, let's say 17.5. This is mapped into the number 306.25. Now take the following sequence of values of x and tabulate their squares: 17.4, 17.41, 17.42, 17.43, 17.44, 17.45, 17.46, 17.47, 17.48, 17.49, 17.491, 17.492, 17.493, 17.494, 17.495, 17.496, 17.497, 17.498, 17.499, 17.4991, 17.4992, 17.4993 (continue for as long as you please). What do you observe? Repeat this procedure from the right-hand side of 17.5 (*i.e.,* start a sequence at 17.6 (say) and approach 17.5 in steps). [You might enjoy playing this game with the other functions on your calculator as well.]

(**b**) Next, consider the function defined by

$$f(x) = \begin{cases} 1 \text{ for } 0 \le x < 1 \\ \\ 2 \text{ for } 1 \le x < 2 \,. \end{cases} \tag{5.1}$$

Draw a graph of this function. Notice that the number 1 is mapped into the number 2. Now take the sequence of values: 0.9, 0.91, 0.92,...,0.999, *etc*. Where does f take them? Discuss. □

The mathematical property exhibited in Experiment 14 by the function $f(x) = x^2$ is *continuity*: as the value 17.5 is approached from the left and from the right, the corresponding values of the function get ever closer to the value $f(17.5) = 306.25$. In contrast, the second function in Experiment 14 suffers a *discontinuity* at the point $x = 1$.

We will now present a more precise discussion of continuity of a function. In doing this, we make frequent use of the mathematical device of intersecting a set with a closed ball, which was introduced in Chapter 3. (Open balls can also be used.) Let $\mathcal{S}$ and $\mathcal{T}$ be nonempty sets of points and let f map $\mathcal{S}$ into $\mathcal{T}$. We will suppose that both $\mathcal{S}$ and $\mathcal{T}$ are subsets of three-dimensional space, but allow for the possibility that one (or both) might in fact be contained in a plane or a line, for example. Choose any element x_0 of $\mathcal{S}$; its image $f(x_0)$ lies in $\mathcal{T}$ (Fig. 24). Let us intersect the set $\mathcal{T}$ with a closed ball $\overline{\mathcal{B}}_\varepsilon(f(x_0))$ of some radius $\varepsilon > 0$, centered at $f(x_0)$. For brevity, we call the ball $\overline{\mathcal{B}}_\varepsilon(f(x_0))$ an ε-ball. Note that the intersection of the ε-ball and the set $\mathcal{T}$ is not necessarily itself a ball, but is certainly always a subset of one. The number ε may be large or small at this stage, but we will be primarily interested in what happens as ε approaches zero. Suppose now that we can find a closed ball $\overline{\mathcal{B}}_\delta(x_0)$, of some radius $\delta > 0$ and centered at x_0, such that the points that lie in the intersection of this δ-ball and $\mathcal{S}$ (*i.e.*, $\overline{\mathcal{B}}_\delta(x_0) \cap \mathcal{S}$) are all mapped into points contained in the ε-ball. (Other points in $\mathcal{S}$ may also be mapped into the ε-ball.) Suppose, moreover, that no matter what size ε is (with the emphasis on smallness), there is always a choice of δ which ensures that all the points in $\overline{\mathcal{B}}_\delta(x_0) \cap \mathcal{S}$ are mapped into points in $\overline{\mathcal{B}}_\varepsilon(f(x_0) \cap \mathcal{T}$. Then, f is said to be *continuous at the point* x_0.

A function may be continuous at all, some, or none of the points of its domain. The functions $f(x) = x^2$, $f(x) = e^x$ and the trigonometric

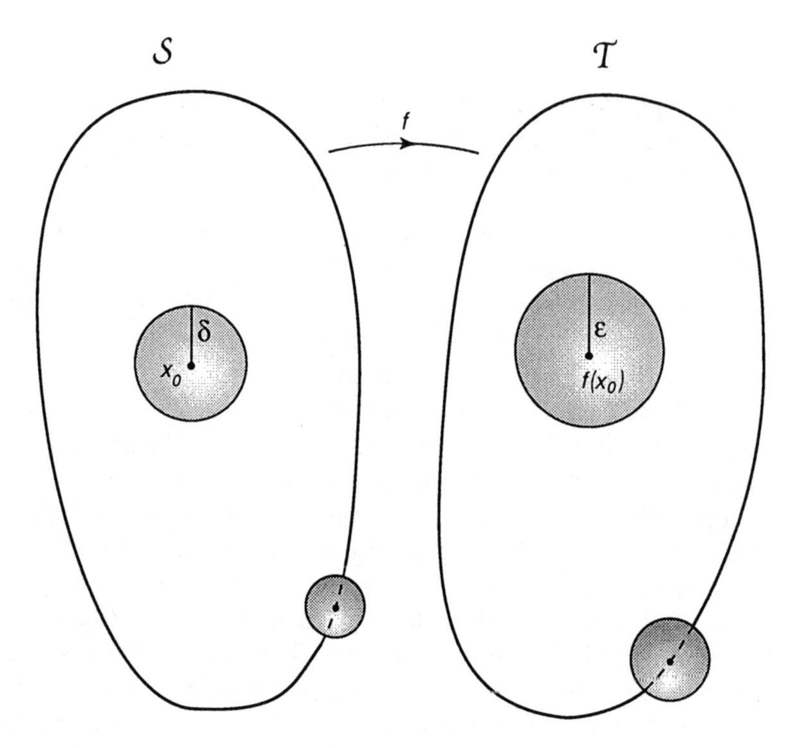

Figure 24 Defining continuity of a function f at a point x_0

functions sin x, cos x, tan x are continuous at every real number. The function defined on the domain of real numbers by $f(x) = 1$ if x is a rational number and $f(x) = 0$ if x is an irrational number is not continuous at any point of its domain.

In Chapter 2, the notion of topological equivalence of geometrical figures was introduced. Using the machinery developed in the present chapter, we can now formally define a *topological mapping* $f : \mathcal{A} \to \mathcal{B}$ as any mapping that satisfies all of the following conditions:

(1) f is continuous at every point of its domain $\mathcal{A}$;

(2) f is bijective (and therefore possesses an inverse $f^{-1} : \mathcal{B} \to \mathcal{A}$);

(3) f^{-1} is continuous at every point of *its* domain, the set $\mathcal{B}$.

A topological mapping is also called a *homeomorphism*. Two sets $\mathcal{A}, \mathcal{B}$ are topologically equivalent if and only if there exists a topological mapping that maps one of them into the other.

Example 1. In a plane, choose a closed line segment MN of unit length and a point O not on the segment (Fig. 25). Join O to the ends of the segment by straight lines OP, OQ. Draw any line parallel to MN and meeting OP and OQ in A and B, respectively. Also, with center O and any radius (> 0), draw an arc of a circle intersecting OP and OQ in C and D, respectively. Finally, pick any point X on the unit segment, join OX and extend this line to intersect AB and CD in

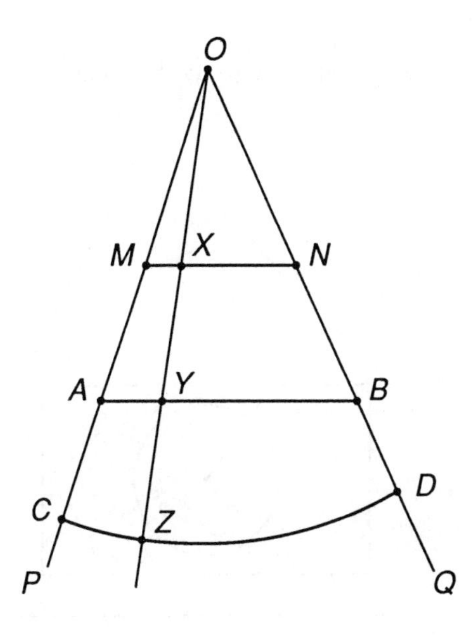

Figure 25 Topological mappings from a unit line segment MN to a line segment and to an arc of a circle

Y and Z, respectively. We now have a way of setting up a one-to-one correspondence between the closed unit segment and the closed segment AB, and, additionally, between the unit segment and the arc CD (with endpoints included). This mapping satisfies the conditions (1), (2), (3) for a topological mapping. It demonstrates that all closed line segments are topologically equivalent to one another, and that circular arcs such as CD are also topologically equivalent to closed line segments.

Example 2. In the plane of Fig. 26, a circle, a rectangle, and a triangle are drawn and a radius OX of the circle is extended to intersect the rectangle and triangle in Y and Z, respectively. As X is moved around the circle, Y traverses the sides of the rectangle, while Z moves around the sides of the triangle. Here, we have a topological mapping which shows that circles, squares, triangles (and other shapes that you can add in a similar manner to Fig. 26) are topologically equivalent to one another.

Example 3. Consider a hemisphere, with center A, resting on a hori-

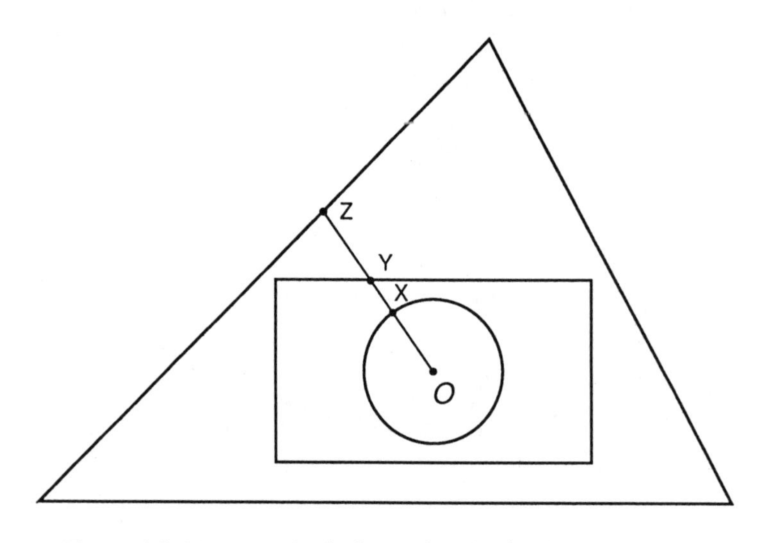

Figure 26 More topological mappings in the plane

zontal plane at B (Fig. 27). Choose a point O located above A on the vertical line joining B and A. Let X be any point on the hemispherical surface (including its edge). Join OX and produce this line until it

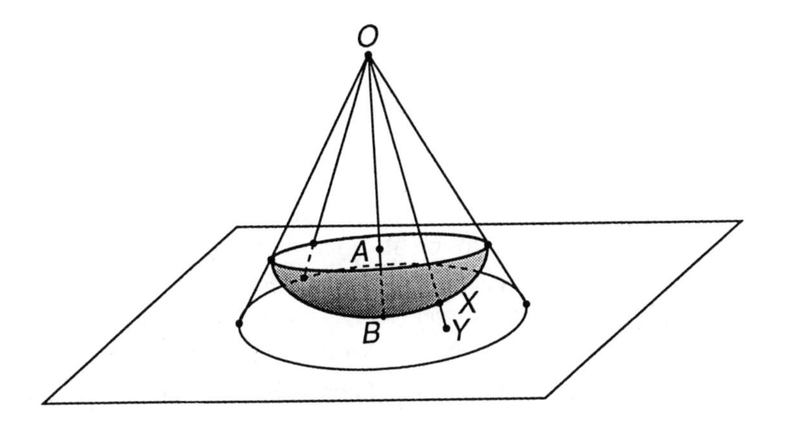

Figure 27 A topological mapping from a hemisphere to a disk

intersects the horizontal plane at Y. As X roams over the hemisphere, its image Y fills up a closed disk on the horizontal plane. We now have a topological mapping which shows that hemispheres are topologically equivalent to closed disks.

Example 4. Consider the following mapping whose domain is a closed disk of radius one:

$$x = \lambda_1 X \quad , \quad y = \lambda_2 Y \qquad (\lambda_1 > 0, \; \lambda_2 > 0) \; . \tag{5.2}$$

Here, (X, Y) are the coordinates of a point on the disk and (x, y) are the coordinates of the image of that point; λ_1 and λ_2 are positive constants, representing stretches (recall Experiment 5). Points lying on the circumference of the unit disk satisfy the equation

$$X^2 + Y^2 = 1 \; . \tag{5.3}$$

But, $X = x/\lambda_1$ and $Y = y/\lambda_1$. Hence, the images of points on the circumference of the disk satisfy the equation

$$\frac{x^2}{\lambda_1^2} + \frac{y^2}{\lambda_2^2} = 1 \;, \tag{5.4}$$

which describes an ellipse with semi-axes λ_1 and λ_2. Similarly, smaller circles are mapped into smaller ellipses. The origin $(X = 0, Y = 0)$ remains fixed. The disk is mapped homeomorphically (or topologically) into an elliptical region.

Example 5. Consider the function defined on the open interval $-\pi/2 < x < \pi/2$ by $y = \tan x$. y takes on all values between $-\infty$ and ∞. The function is bijective and continuous, and its inverse is continuous. It establishes that the open interval $-\pi/2 < x < \pi/2$ is topologically equivalent to the whole real line. In a similar way, it can be shown that every open interval is topologically equivalent to the real line.

From Example 1, it is clear that every interval $a \leq x \leq b$ is topologically equivalent to the closed interval $0 \leq x \leq 1$. Likewise, from Example 5, it follows that every open interval $a < x < b$ is topologically equivalent to the open interval $0 < x < 1$. However, open intervals are never topologically equivalent to closed intervals. In particular, the interval $0 < x < 1$ cannot be deformed into $0 \leq x \leq 1$. Also, the circle is not topologically equivalent to any line segment.

It may seem obvious intuitively, but it is actually difficult to prove that a line segment and a square cannot be topologically equivalent, and likewise for a square and a cube.

Experiment 15 (A family of shapes): Take a rubber band and stretch it into circular, square, and triangular shapes like those in Fig. 26. Form several other shapes as well. Can you deform the band between one shape and another in this family? □

Each one of the figures you formed in Experiment 15 represents a single topological mapping from the unstretched shape of the rubber band.

Altogether, you have a sequence of topological mappings. Moreover, it is clear that by carefully controlling the way in which you pull the rubber band, you can make a sequence of shapes that evolve gradually from one another. Now, we already know that stretching is an example of a deformation. What then is the exact relationship between deformations and topological mappings? We could say that a deformation is a sequence of topological mappings that change gradually from one to another. Let us pursue this idea a little further.

Experiment 16 (A family of mappings): You will need the same apparatus that you used in Experiment 5 and a watch that reads out seconds. First, observe how the rectangle (or circle) gradually changes its shape as you stretch the elastic sheet. Practice performing the process both fast and slowly. Take, let's say, 20 seconds to perform the stretch at a uniform rate. Is it now not true that the longitudinal stretch λ_1 and the transverse stretch λ_2 can be written as $\lambda_1 = c_1 t$ and $\lambda_2 = c_2 t$, where c_1 and c_2 are constants? Evaluate c_1 and c_2. □

If the stretching in Experiment 16 is performed at an unsteady rate, the stretches λ_1 and λ_2 will no longer be proportional to t, but they will still be functions of t: we may then write the deformation as

$$x = \lambda_1(t) X \ , \quad y = \lambda_2(t) Y \ , \tag{5.5}$$

where (X, Y) are the rectangular Cartesian coordinates of an arbitrary point on the unstretched strip and x, y are coordinates of the same point in the stretched strip at time t. We take $\lambda_1(0) = 1$, $\lambda_2(0) = 1$, corresponding to the strip being unstretched at the moment we begin our time-measurements. The pair of Equations (5.5) describe a family of topological mappings, one mapping for each t. The family of mappings is described using time t as a *parameter*, or more briefly, the family is *parametrized by* t. If the functions $\lambda_1(t)$ and $\lambda_2(t)$ are continuous functions of time, as they are in the experiment, we have a *continuous family* of topological mappings. This brings us to a good definition of a deformation: *A deformation is any continuous family of topological mappings parametrized by time (or any other convenient parameter).*

It is essential that the dependence on the parameter be continuous: it serves to exclude from the family topological mappings such as the perfect cut-and-join one that we met in Experiment 8.

Not all deformations are as simple as the one you performed in Experiment 16. For example, if you got someone to help you, you could arrange to have quite different stretches in one portion of the elastic strip than in another, with a gradual transition in between. Such a deformation would be described by equations of the form

$$x = f_1(x, t) \ , \quad y = f_2(x, t) \ . \tag{5.6}$$

These are more complicated than Equations (5.5), but the functions f_1 and f_2 can still be estimated from experimental measurements.

6

Keeping Track of Magnitude, Direction and Sense: Vectors

In addition to possessing magnitude, many familiar physical quantities, such as wind velocity and force, involve direction in an essential way. These quantities can be represented mathematically by *vectors*. It turns out that vectors are also very useful in geometry, and you will have ample opportunity to see this when we study curves and surfaces. In the present chapter, the basic properties of vectors are described.

The concept of vector is a mathematical invention that has its roots in concrete situations. Suppose, for example, that we wish to specify the position of another point B relative to another point A in space (Fig. 28). One way of doing this is to say that B

(i) lies on a line through A parallel to some known line such as CD– this specifies a *direction*;

(ii) is to be reached by proceeding from A in the same sense as from C to D–this specifies a *sense* (or *orientation*); and

(iii) is a certain number of units of length along the line AB–this specifies a *magnitude*.

Thus, the information that we need can be provided by specifying the direction, sense, and magnitude of a line segment, and is independent of the position in space of the line segment. In other words, all other line segments that are parallel to CD, equal to CD in magnitude, and have the same sense as CD, would carry the same three pieces (i), (ii), (iii) of information. Just as numbers were invented through a process of

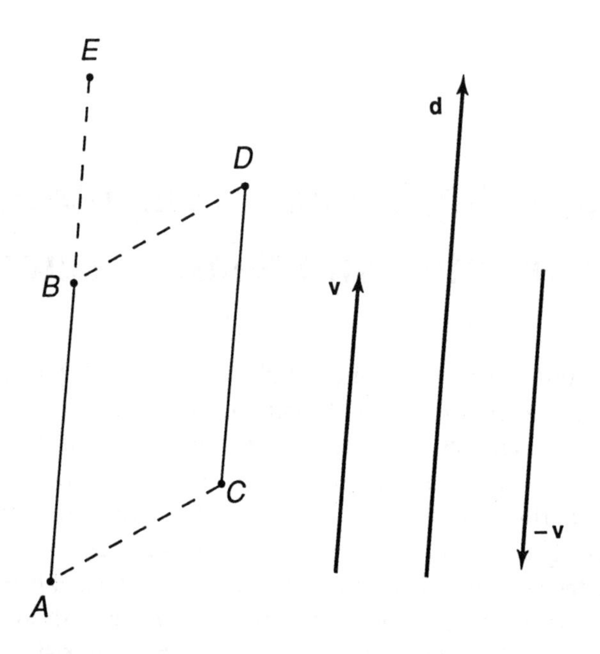

Figure 28 Representing a vector by an arrow

abstraction from concrete objects, we may abstract from line segments and other objects the properties of direction, sense, and magnitude to build a set of new mathematical entities, called vectors, that are ascribed these properties. A (nonzero) vector may be represented visually as an oriented (or "directed") straight-line segment, or "arrow" (Fig. 28): the length of the segment represents the magnitude of the vector; the direction of the segment represents the direction of the vector (in the meaning of (i) above); the sense is represented by the arrowhead. (Often, the word direction is used to include sense as well.) Symbolically, we denote vectors by boldface letters, such as **v** in Fig. 28. The notation $||\mathbf{v}||$ stands for the magnitude of **v**.

If we let the point B in Fig. 28 approach A, then the vector **v** becomes smaller and smaller, but keeps its original direction and sense. In the limit, we obtain an object that has zero magnitude and whatever direction and sense we started out with. This object is the zero vector, denoted by $\mathbf{0}$; its magnitude is zero, and its direction and sense are arbitrary.

As with numbers, mathematical operations can be defined on vectors and these also are suggested by concrete considerations. Thus, consider a point E lying on an extension of AB (Fig. 28). Let the length of the segment AE be c times the length of AB. Then, the position of E relative to A could be represented by a vector $\mathbf{d}$, having the same direction and sense as the vector $\mathbf{v}$ in Fig. 28, and a magnitude that is c times the magnitude of $\mathbf{v}$. Let us adopt as a *definition* that $\mathbf{d}$ is c times $\mathbf{v}$: $\mathbf{d} = c\,\mathbf{v}$. For any vector $\mathbf{v}$, $0\,\mathbf{v} = \mathbf{0}$.

If the point B were located at the same distance from A as in Fig. 28, and also on a line parallel to CD, but now in the sense of D to C, its position relative to A would be represented by a vector having the same magnitude as $\mathbf{v}$ and parallel to $\mathbf{v}$, but with the arrowhead at the opposite end: we may denote this vector by $-\mathbf{v}$. The vector $-(c\,\mathbf{v})$ would then represent the position of E relative to A, if E were along AB and at the same distance as it is in Fig. 28, but now on the opposite side of A; this would also be represented by the vector $c(-\mathbf{v})$.

Next, how are we to combine vectors with one another? For a clue to a good answer, take the same two points A, B of Fig. 28 and suppose we have another point G whose position relative to B is represented by a vector $\mathbf{u}$ (Fig. 29). Clearly, the position of G relative to A is represented by a vector $\mathbf{w}$ which is parallel to the line segment AG, has a magnitude equal to the length of AG and is oriented from A to G. Placing the tail of the vector $\mathbf{u}$ at the tip of the vector $\mathbf{v}$, and placing the tail of $\mathbf{w}$ at the tail of $\mathbf{v}$, we find that the tip of $\mathbf{w}$ coincides with the tip of $\mathbf{u}$ (Fig. 29b). As a *definition* of vector addition, we take: $\mathbf{v} + \mathbf{u} = \mathbf{w}$. This is called the *Triangle Law*. If we complete the parallelogram $ABGH$ in Fig. 29a and also draw copies of $\mathbf{u}$ and $\mathbf{v}$ in Fig. 29b to complete the parallelogram there, $\mathbf{w}$ now appears as the diagonal of the latter parallelogram. For this reason, the operation of vector addition is said to satisfy the *Parallelogram Law*. Note that this construction yields the algebraic rule $\mathbf{u} + \mathbf{v} = \mathbf{v} + \mathbf{u}$ (commutativity).

For vectors, the operation of subtraction may be defined simply as

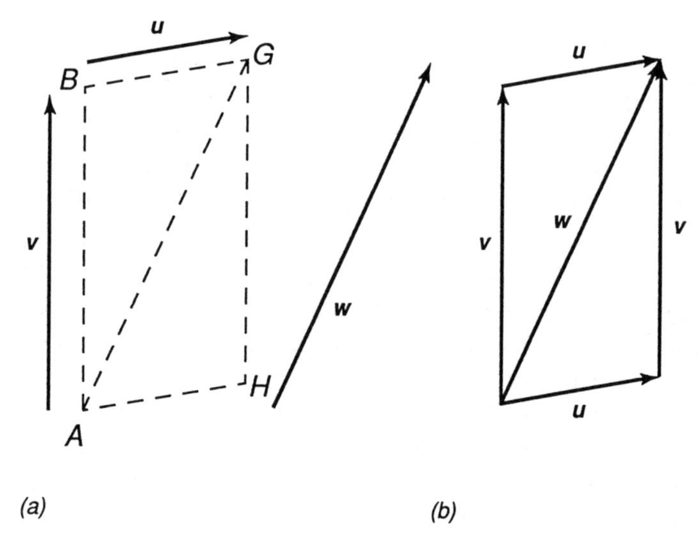

Figure 29 Adding vectors

$\mathbf{u} - \mathbf{v} = \mathbf{u} + (-\mathbf{v})$. For any vector $\mathbf{u}$, $\mathbf{u} - \mathbf{u} = \mathbf{0}$.

Example 1: Position vector. Referring back to the rectangular coordinate system sketched in Fig. 13, let us place a vector $\mathbf{i}$ of unit length joining the origin O to the point I (Fig. 30). Likewise, place a unit vector $\mathbf{j}$ along OJ, and a unit vector $\mathbf{k}$ along OK. The three vectors $\mathbf{i}, \mathbf{j}, \mathbf{k}$ are said to form an orthonormal triad–"normal" in the sense of having unit magnitude, and "ortho" abbreviating orthogonal (or perpendicular), since in each pair $(\mathbf{i}, \mathbf{j})$, $(\mathbf{j}, \mathbf{k})$, $(\mathbf{k}, \mathbf{i})$, the two vectors are mutually perpendicular.

The vector $\mathbf{r}$ joining O to a generic point A, with coordinates (x, y, z), is called the *position vector* of A. The vector $x\mathbf{i}$ takes us out to M, $y\mathbf{j}$ takes us out to N, and $z\mathbf{k}$ takes us out to the point W on the z-axis that is at the same height as A. The vector sum $x\mathbf{i} + y\mathbf{j}$ takes us from O to B, and then adding $z\mathbf{k}$, we are taken from B to A. Thus, the position vector $\mathbf{r}$ has the representation

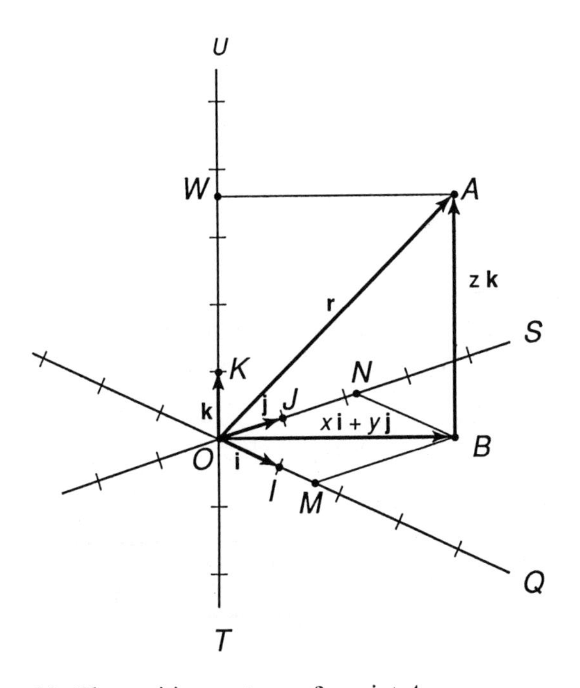

Figure 30 The position vector $\mathbf{r}$ of a point A

$$\mathbf{r} = x\mathbf{i} + y\mathbf{j} + z\mathbf{k} \ . \tag{6.1}$$

The three vectors $x\mathbf{i}$, $y\mathbf{j}$, and $z\mathbf{k}$ are the *rectangular components* of $\mathbf{r}$. The zero vector can be written as $\mathbf{0} = 0\mathbf{i} + 0\mathbf{j} + 0\mathbf{k}$.

The magnitude of $\mathbf{r}$ is equal to the distance from O to A. Hence, by Pythagoras's Theorem,

$$\|\mathbf{r}\| = \sqrt{x^2 + y^2 + z^2} \ . \tag{6.2}$$

Given an arbitrary nonzero vector $\mathbf{v}$, we may place the tail of $\mathbf{v}$ at O. Its tip will then locate some unique point in space, and we may proceed as we did for $\mathbf{r}$ to decompose $\mathbf{v}$ into components lying along the three coordinate axes. We write

$$\mathbf{v} = v_x \, \mathbf{i} + v_y \, \mathbf{j} + v_z \, \mathbf{k} \, . \tag{6.3}$$

The magnitude of **v** is given by

$$\|\mathbf{v}\| = \sqrt{v_x^2 + v_y^2 + v_z^2} \, . \tag{6.4}$$

Example 2: Spinning wheel. Consider a circular disk that is spinning at a constant rate on a horizontal axle through its center O (Fig. 31). Let its period of rotation be T seconds, and suppose that it is spinning in

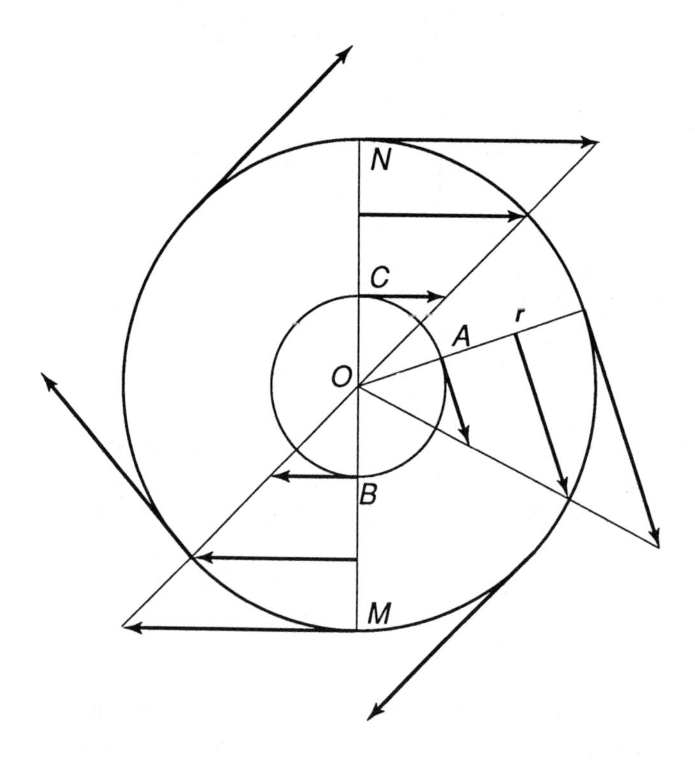

Figure 31 Velocity of points on a spinning wheel

the clockwise direction. We would like to draw vectors representing the velocity of various points on the wheel.

The velocity of O is the zero vector $\mathbf{0}$. For any particle A other than O, as the disk spins, A moves in a circle ABC; its velocity vector $\mathbf{v}$ is tangent to this circle. The magnitude $\|\mathbf{v}\|$ of $\mathbf{v}$ is the speed at which A moves. Each particle on the circle ABC has the same speed, but a different velocity. Particles lying on larger circles (centered at O) have greater speeds. We may calculate the speed of A as follows: since A travels one revolution ($= 2\pi$ radians) in T seconds, it covers a distance $2\pi r$ (meters, say) in this time. Hence, its speed is $2\pi r/T$ meters per second. The quantity $2\pi/T$ is called the angular velocity of the disk, measured in radians per second, and is usually denoted by the letter ω. Hence, the speed of A is $r\omega$: it is a linear function of r. Let us draw a vertical diameter MN of the disk. The velocity of each particle on MN is given by a horizontal vector whose length is proportional to the distance of the particle from O, and whose sense is to the right for particles located above O and to the left for particles below O. The velocity vectors of various other points are also indicated in Fig. 31.

Experiment 17 (Bicycle wheel): Suppose that you are riding a bicycle at constant speed along a straight level road. The front wheel certainly has a definite angular velocity. However, the center of the wheel is not at rest, but has the same velocity as the frame of the bicycle. Reflecting on the fact that if you ride through a puddle, the wheel will produce a clearly defined track, argue that the velocity of the particle at the bottom of the wheel must be zero at the moment it is in contact with the road. (Note: if you apply your brakes, the track would be smeared; correspondingly, the velocity of the contacting particle would not then be zero.) Deduce that the velocity vectors of particles on the bicycle wheel can be obtained by adding vectorially the constant forward velocity vector of the bicycle to the velocity vectors of the spinning wheel in Example 2. Carry out this operation accurately on a sheet of paper. Convince yourself that each particle on the wheel can be considered as moving instantaneously in a circle centered at the point of contact with the road. □

Example 3: Vector equation of a line. Consider a line drawn through two distinct points M and N in space (Fig. 32). Let $\mathbf{r}_M$ and $\mathbf{r}_N$ be the position vectors of M and N. The vector $\mathbf{r}_N - \mathbf{r}_M$ takes us from M to N : $\mathbf{r}_M + (\mathbf{r}_N - \mathbf{r}_M) = \mathbf{r}_N$. Dividing $\mathbf{r}_M - \mathbf{r}_N$ by its own magnitude, we get a unit vector $\mathbf{u}$ that is parallel to the given line and has the sense indicated in Fig. 32. Now take a generic point A belonging to the line, and let ξ be the distance from M to A if A lies to the right of M, and the negative of the distance if A lies to the left of M. The position vector of A can then be written as

$$\mathbf{r} = \mathbf{r}_M + \xi \mathbf{u} \ . \tag{6.5}$$

Let us decompose each of the vectors $\mathbf{r}$, $\mathbf{r}_M$, and $\mathbf{u}$ into its rectangular components:

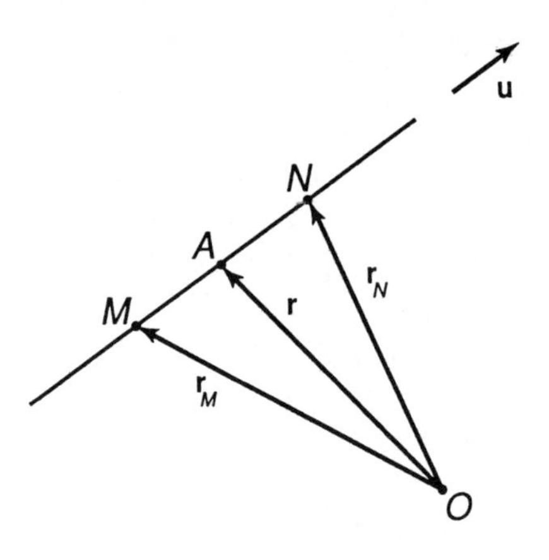

Figure 32 Obtaining the vector equation of a line

$$\begin{aligned}
\mathbf{r} &= x\mathbf{i} + y\mathbf{j} + z\mathbf{k} \ , \\
\mathbf{r}_M &= x_M\,\mathbf{i} + y_M\,\mathbf{j} + z_M\,\mathbf{k} \ , \\
\mathbf{u} &= u_x\,\mathbf{i} + u_y\,\mathbf{j} + u_z\,\mathbf{k} \ .
\end{aligned} \tag{6.6}$$

Inserting Equations (6.6) into Equation (6.7), we obtain

$$(x - x_M - \xi\, u_x)\mathbf{i} + (y - y_M - \xi\, u_y)\mathbf{j} + (z - z_M - \xi\, u_z)\mathbf{k} = \mathbf{0} \ . \tag{6.7}$$

Since a vector equals zero only if its magnitude is zero, and hence only if each component of it is zero, it follows from Equation (6.7) that

$$x = x_M + \xi\, u_x \ , \quad y = y_M + \xi\, u_y \ , \quad z = z_M + \xi\, u_z \ . \tag{6.8}$$

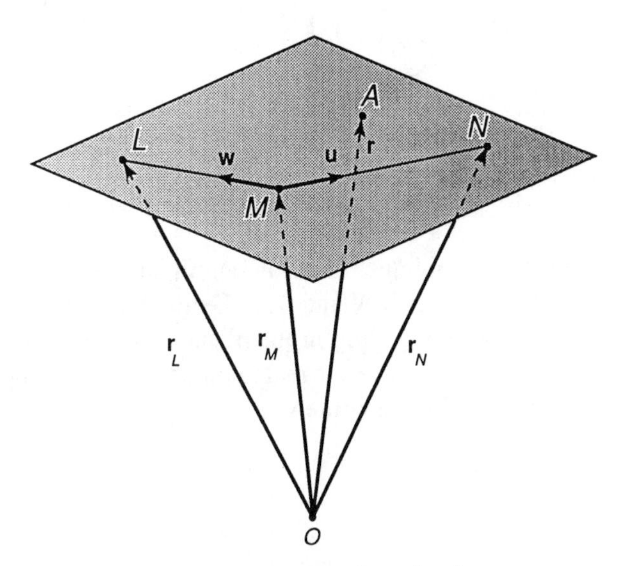

Figure 33 Obtaining the vector equation of a plane

The three coordinate Equations (6.8) are together equivalent to the vector Equation (6.5).

Example 4: Equation of a plane. Let L, M, N be three noncollinear points, and pass a plane through them (Fig. 33). Denote the position vectors of L, M, N by $\mathbf{r}_L, \mathbf{r}_M, \mathbf{r}_N$, and let $\mathbf{r}$ be the position vector of an arbitrary point A on the plane. As in Example 2, we can construct a unit vector $\mathbf{u}$ by dividing $\mathbf{r}_N - \mathbf{r}_M$ by $||\mathbf{r}_N - \mathbf{r}_M||$; similarly, construct a unit vector $\mathbf{w}$ by dividing $\mathbf{r}_L - \mathbf{r}_M$ by $||\mathbf{r}_L - \mathbf{r}_M||$. The vectors $\mathbf{u}, \mathbf{w}$ can be placed with their tails at M: $\mathbf{u}$ then lies along MN and points towards N, while $\mathbf{w}$ lies along ML and points towards L. It is convenient to set

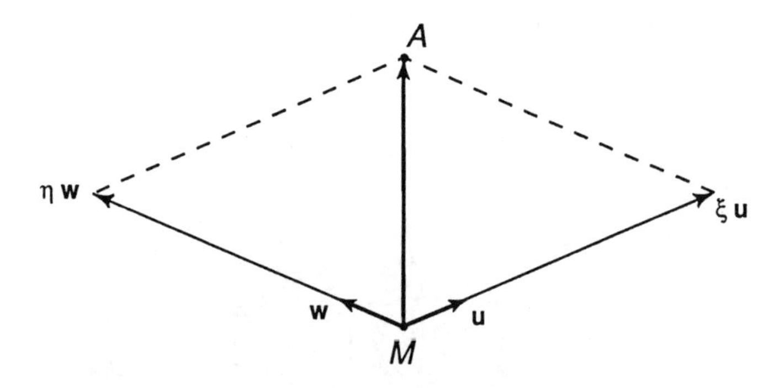

Figure 34 Oblique Cartesian coordinate system in a plane

up a planar system of oblique Cartesian coordinates, with M as origin and with axes lying along MN and ML. Using the parallelogram law, we can express any vector lying in the plane as a sum of a vector $\xi\mathbf{u}$ and a vector $\eta\,\mathbf{w}$, where ξ and η are real numbers (Fig. 34). The position vector of A may therefore be written as

$$\mathbf{r} = \mathbf{r}_M + \xi\,\mathbf{u} + \eta\,\mathbf{w} \ . \tag{6.9}$$

As in (6.8), three coordinate equations can be derived from Equation (6.9). These will now involve two parameters (ξ and η).

It is convenient to introduce here a special operation on vectors that will prove useful in later chapters. It is called the *dot product* (or *scalar product*) and is defined as follows. Let **u** and **v** be any vectors. The dot product of **u** and **v** is denoted by $\mathbf{u} \cdot \mathbf{v}$ and is a real number assigned according to the rules:

(i) if **u** or **v** is the zero vector, then

$$\mathbf{u} \cdot \mathbf{v} = 0\;; \tag{6.10}$$

(ii) if both **u** and **v** are nonzero, and α $(0^\circ \leq \alpha \leq 180^\circ)$ is the angle between them (Fig. 35), then

$$\mathbf{u} \cdot \mathbf{v} = \|\mathbf{u}\|\, \|\mathbf{v}\| \cos\alpha \quad . \tag{6.11}$$

The right-hand side of Equation (6.11) can be regarded as the magnitude of **u** multiplied by the magnitude $\|\mathbf{v}\| \cos\alpha$ of the perpendicular projection of **v** onto **u**, and also as the magnitude of **v** multiplied by the magnitude $\|\mathbf{u}\| \cos\alpha$ of the perpendicular projection of **u** onto **v** (Fig. 35).

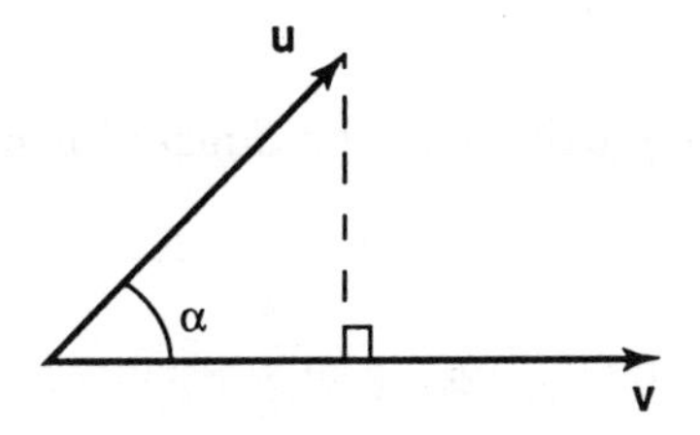

Figure 35 Defining the dot product of **u** and **v**

The dot product has the following properties:

$$\begin{aligned} &(a) \quad \mathbf{u} \cdot \mathbf{v} = \mathbf{v} \cdot \mathbf{u} \\ &(b) \quad \mathbf{u} \cdot (c\mathbf{v}) = c(\mathbf{u} \cdot \mathbf{v}), \quad (c = \text{ any real number}) \\ &(c) \quad \mathbf{u} \cdot (\mathbf{v} + \mathbf{w}) = \mathbf{u} \cdot \mathbf{v} + \mathbf{u} \cdot \mathbf{w} \ . \end{aligned} \tag{6.12}$$

Properties (a) and (c) are called the commutative and distributive laws, respectively.

If $\mathbf{v}$ is chosen equal to $\mathbf{u}$ ($\mathbf{u} \neq \mathbf{0}$) in Equation (6.11), we obtain

$$\mathbf{u} \cdot \mathbf{u} = ||\mathbf{u}||^2 \ , \tag{6.13}$$

since the angle α is $0°$ for this case. Equation (6.13) is also true for the case $\mathbf{u} = \mathbf{0}$, since both sides of the equation are then 0. If two (nonzero) vectors are perpendicular to one another, then $\alpha = 90°$, and hence $\mathbf{u} \cdot \mathbf{v} = 0$. For the unit vectors **i, j, k,** it is evident that

$$\begin{aligned} &\mathbf{i} \cdot \mathbf{i} = \mathbf{j} \cdot \mathbf{j} = \mathbf{k} \cdot \mathbf{k} = 1 \ , \\ &\mathbf{i} \cdot \mathbf{j} = \mathbf{j} \cdot \mathbf{k} = \mathbf{k} \cdot \mathbf{i} = 0 \ . \end{aligned} \tag{6.14}$$

The rectangular Cartesian components of a vector $\mathbf{v}$ (see Equation (6.3)) satisfy the relations

$$\mathbf{v} \cdot \mathbf{i} = v_x, \quad \mathbf{v} \cdot \mathbf{j} = v_y \ , \mathbf{v} \cdot \mathbf{k} = v_z \ . \tag{6.15}$$

Furthermore, we may write the dot product of $\mathbf{u}$ and $\mathbf{v}$ in terms of their components as:

$$\begin{aligned} \mathbf{u} \cdot \mathbf{v} &= \mathbf{u} \cdot (v_x\mathbf{i} + v_y\mathbf{j} + v_z\mathbf{k}) \\ &= u_x v_x + u_y v_y + u_z v_z \ . \end{aligned} \tag{6.16}$$

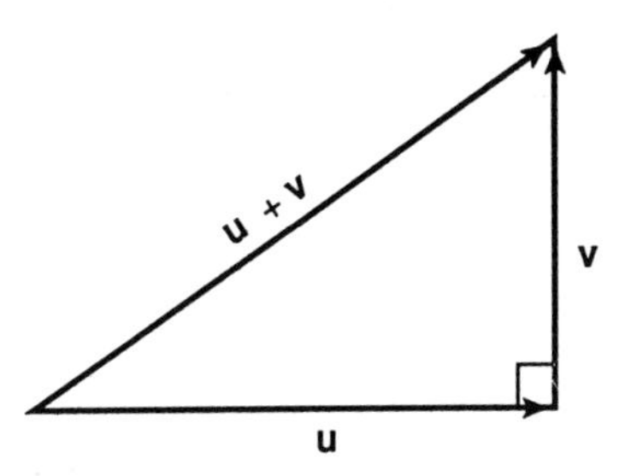

Figure 36 Pythagoras's Theorem anew

With the use of the inner product, we can give a simple algebraic proof of Pythagoras's Theorem. Thus, let $\mathbf{u}$ and $\mathbf{v}$ be two nonzero vectors that are perpendicular to one another. The sum $\mathbf{u} + \mathbf{v}$ is represented by the hypotenuse of the right-angled triangle in Fig. 36. Now, the square of the magnitude of the hypotenuse is given by

$$\begin{aligned} \| \mathbf{u} + \mathbf{v} \|^2 &= (\mathbf{u} + \mathbf{v}) \cdot (\mathbf{u} + \mathbf{v}) \\ &= \mathbf{u} \cdot \mathbf{u} + 2\mathbf{u} \cdot \mathbf{v} + \mathbf{v} \cdot \mathbf{v} . \end{aligned} \tag{6.17}$$

But, since $\mathbf{u}$ and $\mathbf{v}$ are perpendicular to one another, their dot product vanishes. Equation (6.17) therefore reduces to

$$\| \mathbf{u} + \mathbf{v} \|^2 = \| \mathbf{u} \|^2 + \| \mathbf{v} \|^2 . \tag{6.18}$$

7

Curves

In preceding chapters, we have discussed some special types of curves, such as the ellipse, the circle, and the straight line. But, these are only a few of the infinite variety of curves that can be imagined. The path of a bird flying through the air, the instantaneous shape of a swinging chain held at its top, the outlines of petals, the forms of arches and suspension bridges – all provide physical examples of curves. We wish to describe such general curves mathematically and to explore their essential properties. To do so, we will utilize the concepts and theoretical tools that have been introduced in earlier chapters. Our aim is to proceed from intuitive notions about curves to a clear, abstract definition. This process – the clarification of ideas – is really one of the most important activities of the mathematician.

What is a Curve?

Each of us has some idea of what a "curve" is, but to give a precise definition of the concept is not as easy as might first appear. Actually, there are several different (and valuable) ways of thinking about curves, and it will be well worthwhile to sort out some of the differences between them before settling on a final conception.

One fruitful notion, which generalizes an idea that we met in Chapter 4, is that a curve can be regarded as the path traced out by a moving material point. This conception includes curves of the type sketched in Fig. 37, which intersects itself at M and has a loop at N. The direction of travel, or orientation, of the curve is indicated by the arrows placed along it.

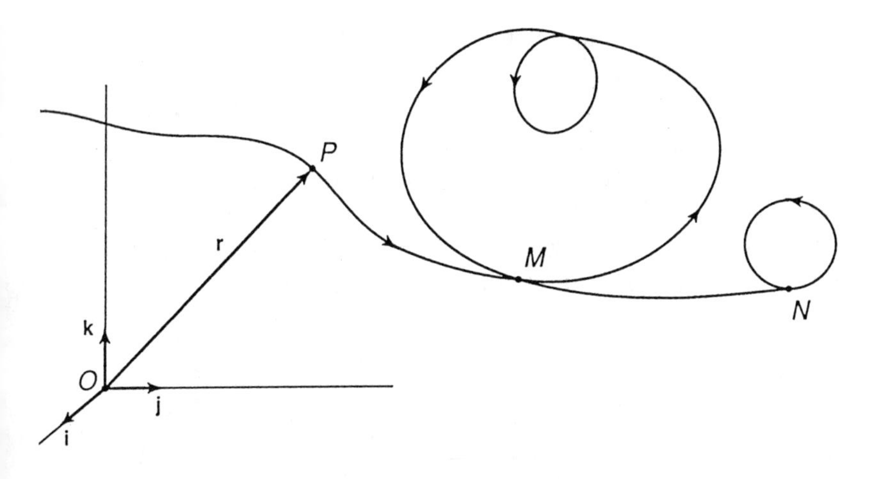

Figure 37 A path traced out by a moving material point

If we translate the above notion of a curve into algebraic language, we may start off by saying that a curve can be regarded as the set of points described by three time-dependent coordinate mappings $\hat{x},\ \hat{y},\ \hat{z}$. These specify the Cartesian coordinates x,y,z of the point P occupied by the moving material point at time t:

$$x = \hat{x}(t)\,, \quad y = \hat{y}(t)\,, \quad z = \hat{z}(t)\,. \tag{7.1}$$

The position vector of P is $\mathbf{r} = x\,\mathbf{i} + y\,\mathbf{j} + z\,\mathbf{k}$, where $\mathbf{i}, \mathbf{j}, \mathbf{k}$ are unit vectors along the coordinate axes. Consequently, we may also specify the curve by the vector expression

$$\begin{aligned} \mathbf{r} &= \hat{x}(t)\,\mathbf{i} + \hat{y}(t)\,\mathbf{j} + \hat{z}(t)\,\mathbf{k} \\ &= \hat{\mathbf{r}}(t) \quad , \end{aligned} \tag{7.2}$$

where the notation $\hat{\mathbf{r}}$ stands for the mapping (or function) that assigns the position vector $\mathbf{r}$ to the instant t.

In our usual understanding of motion, we imagine the variable t to be increasing steadily during some interval $t_1 \leq t \leq t_2$, and that for each value of t in this interval, there is a unique value of $\mathbf{r}$. It is admissible for the same positions to be visited over and over again by the moving point, as for instance in the curve sketched in Fig. 38, or in the case of a planet as it revolves in its elliptical orbit about the sun.

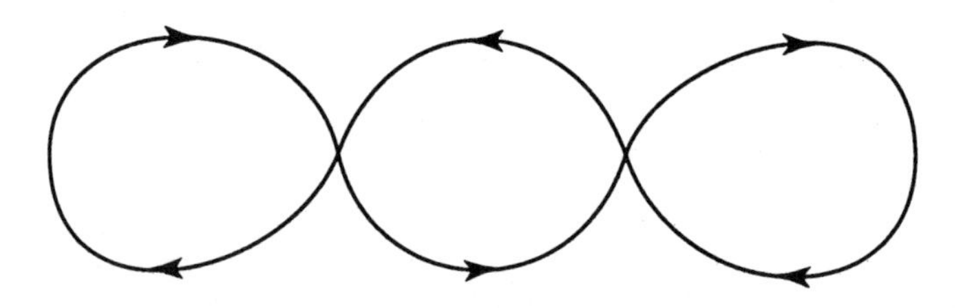

Figure 38 Another path traced out by a moving material point

Different motions can produce the same set of points in space. For example, the same circle is produced whether we let a moving material point revolve through 360° once, or twice, or any number of times, about a center. Likewise, the curve in Fig. 38 can be transited in infinitely many different ways. Indeed, even a unit segment of a straight line can be generated by many different motions. We might, for instance, let the moving point go from the origin to the half-way mark and then back to the origin (Fig. 39); from there, it may proceed three quarters of the way across the interval and again back to the origin; finally, it may go from the origin to the other end of the interval. Of course, the entire segment would have been generated by the motion described in the last

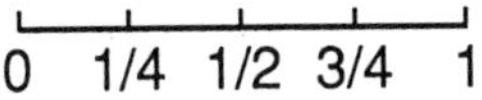

0 1/4 1/2 3/4 1

Figure 39 Moving back and forth along a unit interval

step. The point we are making here is that the way in which a curve is generated by a moving point is far from being unique. While this provides flexibility in the description of curves, it can be somewhat of a nuisance too, because two motions which appear to be very different may in fact generate the same curve.

The state of rest may be regarded as a special motion in which the velocity is always zero. This corresponds to the choice

$$\hat{\mathbf{r}}(t) = \text{constant vector} \tag{7.3}$$

for the function in Equation (7.2). However, we would not regard the specification in Equation (7.3) as describing a curve, because this specification produces only a single point. We may rule out this case by insisting that the moving point not be at rest (except possibly at isolated instants of time).

Example 1. Consider the motion

$$\hat{x}(t) = \frac{1}{2}t^2 - \frac{1}{3}t^3 \ , \quad \hat{y}(t) = 0 \ , \quad \hat{z}(t) = 0 \ , \tag{7.4}$$

with $0 \leq t \leq 1$. The material point moves along a straight-line segment. At $t = 0$, it is at $x = 0$, and at $t = 1$, it is at $x = 1/6$. The velocity of the moving point is found by differentiating the function $\hat{x}$ with respect to t:

$$v = \frac{d\hat{x}}{dt}(t) = t(1 - t) \quad . \tag{7.5}$$

The velocity v is zero at $t = 0$ and again at $t = 1$. Furthermore, v is positive for $0 < t < 1$. Thus, the moving point starts out from rest at the position $x = 0$ at time $t = 0$, and moves to the right without coming to rest until $t = 1$, at which time it occupies the position $x = 1/6$. The graph of the function $\hat{x}$ is plotted in Fig. 40.

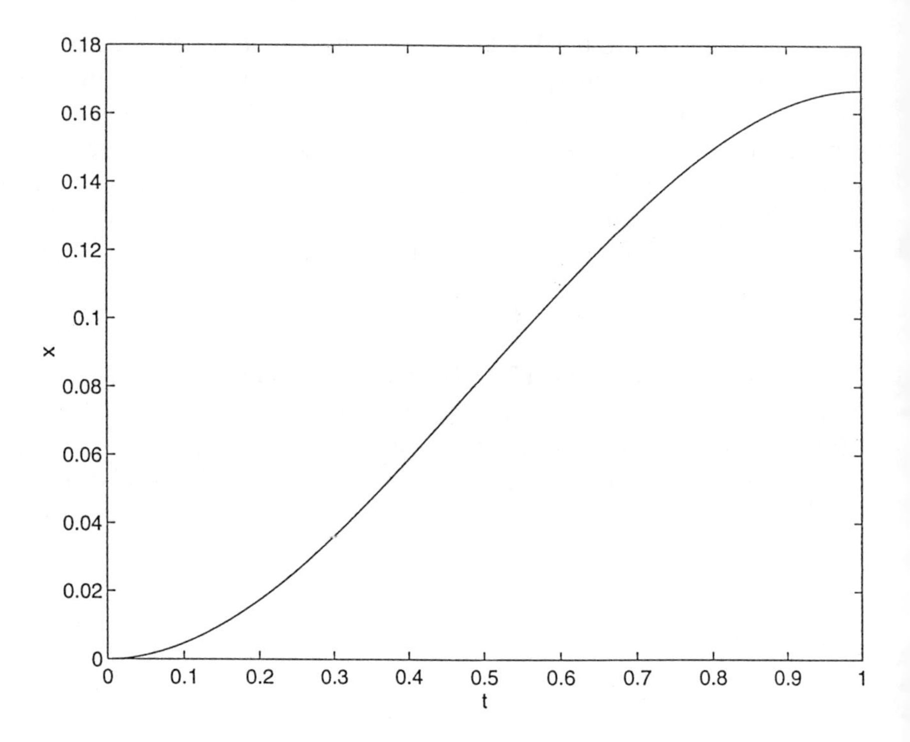

Figure 40 Graph of the function $\hat{x}$

A method of defining a curve that is free from kinematical considerations can be explained with reference to the Example 1. Reconsidering Equation (7.4), we see that to each value of t in the interval $0 \leq t \leq 1$, the equation assigns a unique value of x – as it must do, since $\hat{x}$ is a mapping. Thus, to each point lying between 0 and 1 (including these end-points) on the t-axis in Fig. 40, there corresponds a unique point on the x-axis. The locus of the latter points is the segment $0 \leq x \leq 1/6$. But, it is also true that for each point in the segment $0 \leq x \leq 1/6$, there is a unique point in the segment $0 \leq t \leq 1$. In other words, we have a one-to-one *correspondence* between the two segments. Because of this, we also have at our disposal a mapping in the reverse direction, *i.e.,* from the x-interval into the t-interval. Recall from Chapter 4 that this is called the *inverse* of the mapping $\hat{x}$. Pictorially, we may represent the two functions $\hat{x}$ and its inverse by the diagram in Fig. 41, where the upper segment comprises the values taken on by x, the lower segment

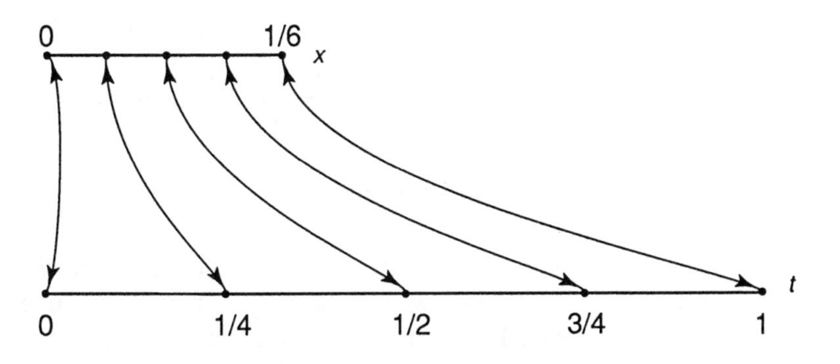

Figure 41 A one-to-one correspondence

the values of t, and the curved arrows indicate corresponding values. For example, t = 1/4 is mapped into x = 5/192, and conversely x = 5/192 is mapped into t = 1/4.

We now come to another significant feature of Example 1. Notice that the points in the segment $0 \leq x \leq 1/6$ correspond to the points in the segment $0 \leq t \leq 1$ not in a haphazard way, but in such a way that points which are close to one another on the t-interval are mapped into points that are close to one another on the x-interval. From the discussion given in Chapter 5, you will recognize that the function $\hat{x}$ is continuous, and so also is its inverse.

If you refer back to the three conditions given on pages 60-61, you will see that the mapping $\hat{x}$ is actually a homeomorphism (or a topological mapping).

In Fig. 42, two other examples are shown in which a segment of a straight line is mapped homeomorphically to produce other sets of points, which we would surely call curves. We are therefore inclined to say that every homeomorphic image of a straight-line segment is a curve.

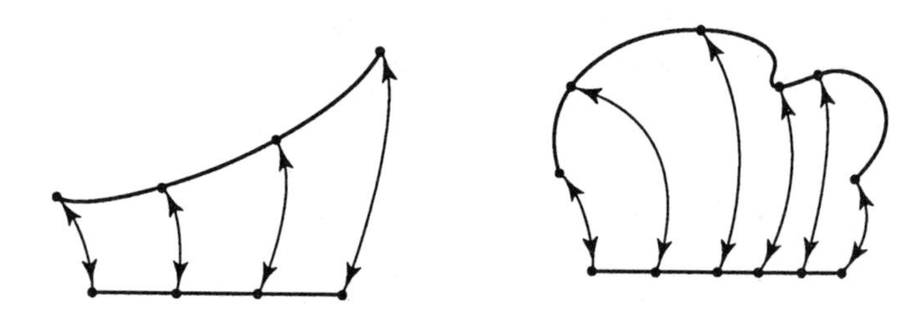

Figure 42 Two more homeomorphisms

We have made a good deal of progress towards an exact conception of "curve". But are all sets that we would wish to regard as curves homeomorphic images of line segments? Consideration of the three examples in Fig. 43 indicates that this question must be answered in the negative. It is not difficult to salvage the first two cases in Fig. 43, because even though neither the pair of disjoint closed line segments nor the circle is a homeomorphic image of a straight-line segment, each of these two sets has the following property: if P is any point of the set, there is some piece QR of the set such that QR contains P and is also the homeomorphic image of a closed straight-line segment (see Fig. 44). Better yet, QR can be obtained through a *deformation* of the closed straight-line segment (recall Experiments 15 and 16). Let us call sets of this type *elementary curves*. With this in mind, we make a formal definition using the device of intersection with a closed ball, which was introduced in Chapter 3:

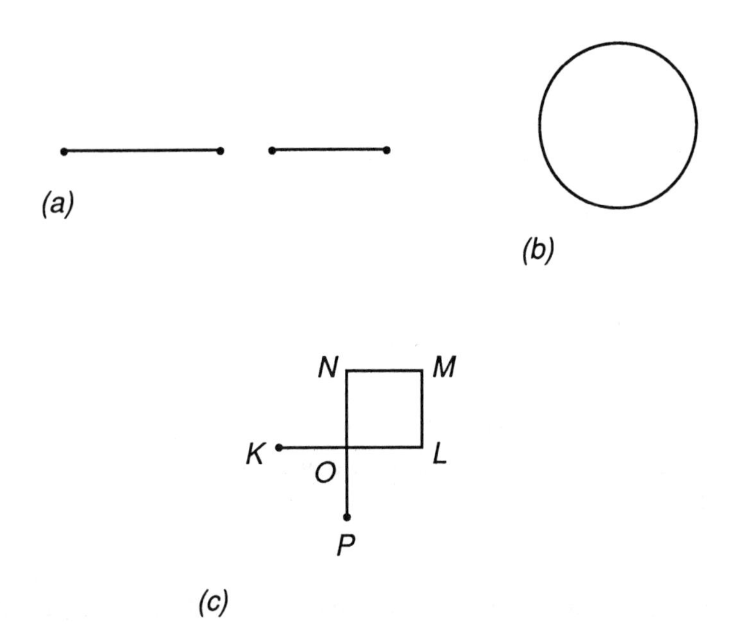

Figure 43 Three objects which cannot be mapped homeomorphically into line segments

Definition (C1): An *elementary curve* is a set $\mathcal{C}$ of points in three-dimensional Euclidean space with the following property: at every point P of $\mathcal{C}$, we can construct a closed ball $\bar{\mathcal{B}}_r(P)$ of radius $r(>0)$ and center P, such that the intersection $\mathcal{N} = \bar{\mathcal{B}}_r(P) \cap \mathcal{C}$ can be obtained by a deformation of some closed straight-line segment.

Thus, in Fig. 45, a closed ball of radius r, centered at P, intersects the given set MN in a segment QR, and the latter can be obtained by deforming a closed straight-line segment $Q'R'$; in this case $\mathcal{N} = QR$. We will call any such piece $QR(\subseteq MN)$ an *arc* of the curve. The arc is really the essence of a curve. All arcs are topologically equivalent to one another. Indeed, they can all be deformed into one another.

It is worth pointing out that if the closed ball in Fig. 45 is centered at

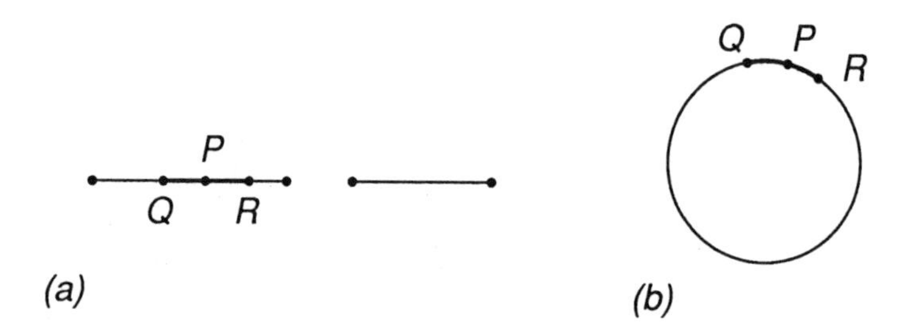

Figure 44 In (a), the segment QR can be deformed into a closed straight-line segment, and likewise for the piece of QR of the circle in (b).

the endpoint N, its intersection with MN would be a closed segment NZ, which again can be obtained by deforming a closed straight-line segment.

It follows from Definition (C1) that every arc of an elementary curve can be parametrized by an expression having the same form as Equation (7.2). But, in general, no one parametrization will work for the whole elementary curve.

In Fig. 46, an open straight-line segment AB (*i.e.,* without end-points) is shown. It was mentioned in Chapter 5 that an open interval can never be homeomorphic to a closed interval. An open interval is therefore not an arc. Nevertheless, at each point P of the open segment AB, we can form the intersection of AB with a small closed ball centered at P. This intersection is itself a closed line segment QR. As P moves closer to where the removed end points A and B were, the diameter QR becomes smaller and smaller, but it still has a nonzero value. We therefore see that the open line segment qualifies as an elementary curve.

What about the third set of points in Fig. 43? At the point of crossing O, a mapping from a straight-line segment would not be one-one, and hence not a homeomorphism. Therefore, the set is not an elementary curve. However, it is a union of two elementary curves, such as $KLMN$ and NOP. Likewise, the set of points in Fig. 37, and also that in Fig. 38, is each a union of a finite number of elementary curves.

The foregoing discussion paves the way for the following general

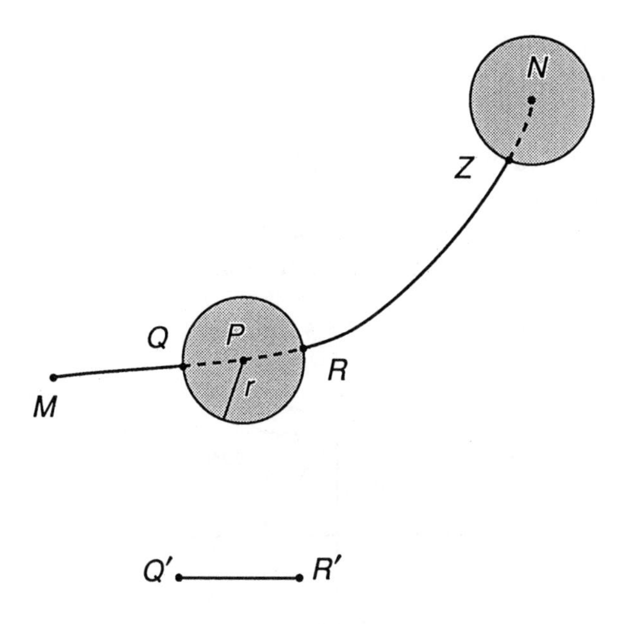

Figure 45 QR can be deformed into the straight-line segment $Q'R'$.

definition:

Definition (C2): A *curve* is any set that can be regarded as the union of finitely many elementary curves.

Another example of a curve that is not an elementary curve is shown in Fig. 47. Usually, there are many ways in which such a curve can be decomposed into elementary curves. It should also be noted that an elementary curve is indeed itself a curve in the sense of Definition (C2).

We have now arrived at a very general conception of "curve". Al-

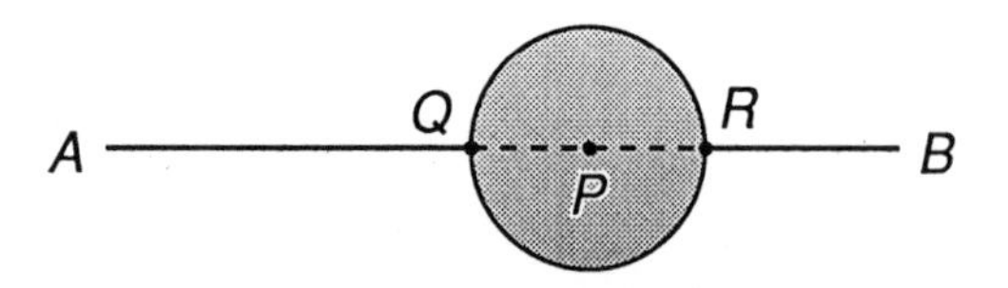

Figure 46 An open interval is an elementary curve

though we started out with the suggestive device of a moving material point, in our final definition no mention of this notion is made. We abstracted the essential content of the concept of "curve" from its intuitive underpinnings. Almost all good definitions in mathematics evolve in this

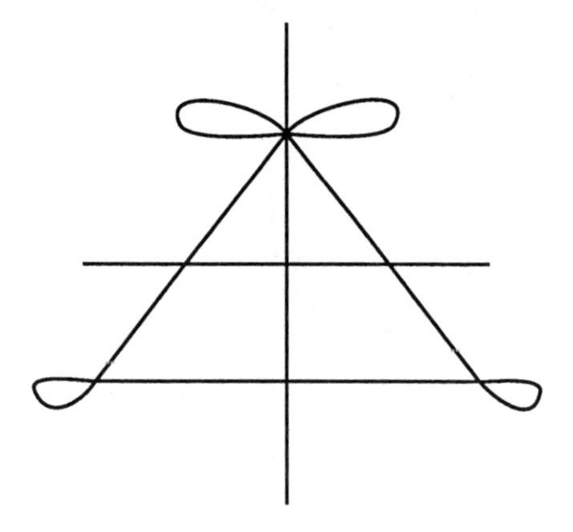

Figure 47 A curve that can be decomposed into many different elementary curves

way, *i.e.,* as abstractions of intuitive notions.[1]

Experiment 18 (Curve test): To determine whether or not a given figure is an elementary curve, one can use the following physical method: Take a short piece of string or a segment of a rubber band and test whether at each point of the given figure the string or rubber segment can be laid on the figure to form a one-to-one correspondence. (Referring to Fig. 45, you can think of the segment QR as being the piece of string or rubber, and of successively laying it along portions of MN.) Try this test out on a variety of figures, including the perimeters of boxes and other objects. □

Experiment 19 (Wires and rubber bands): (**a**) Take a straight piece from fairly stiff wire and bend it into a variety of shapes. Explain why each of these physical shapes corresponds to a mathematical curve.

(**b**) Take a rubber band, stretch it and make it take up a variety of shapes (by having it go around some pegs, for example). Does each such shape correspond to a curve? □

The family of curves which you encountered in Part (b) of Experiment 19 is especially useful in mathematics. They are homeomorphic images of circles, and are called simple (= not self-intersecting) closed curves or *Jordan* curves (some examples appear in Fig. 26). Jordan curves need not be planar, and include knots.

Note to Chapter 7

[1] The concept of a curve can be extended even further to include certain types of union of *infinitely* many elementary curves (see Whyburn (1942)).

8

Arc Length

A geometrical measurement begins physically,
but it is achieved only metaphysically!

H. Lebesgue (1966)

Despite the fundamental importance of the concept of length in Euclidean geometry, the Greek geometers did not know how to define the length of a general curve. In other words, the Greek mathematicians were unable to put into mathematical language an idea that every ancient rope-stretcher and tailor must have known! To understand the nature of the difficulty, let us start with an experiment.

Experiment 20 (Measuring length): (**a**) Use a string (and a ruler), or a measuring tape, to find the length of the curve formed by the bottom of a container, or an oval tray, or your waistband.

(**b**) Measure the length of one of the curves which you made from wire in Experiment 19.

(**c**) Cut a narrow helical groove in the bark of a stick and use a string to measure the length of the groove. □

The properties of the string that we are implicitly appealing to in Experiment 20 are its flexibility and its inextensibility: it can change its shape to fit a given curve, but its length is always the same, and can be found by laying it out straight. When we lay the measuring tape along a given curve, we regard the length of the tape as still being given by the readings on it. Thus, from a practical point of view, we could regard the length of a curve as the number obtained by laying an "inextensible" string along it. But, the meaning of *inextensible* is that the *length* of all possible configurations of the string is the same. This statement makes sense only if we already know how to assign mathematical meaning to the idea of length of a curve. We must take the bull by the horns, as it were, and approach the problem directly.

Definition of Arc Length

Consider an arc AB of a curve (Fig. 48). If we choose some points lying on the arc (such as A_1, A_2, A_3, A_4) between $A(= A_0)$ and $B(= A_5)$ and join A to A_1, A_1 to A_2, *etc.*, we obtain a broken line which

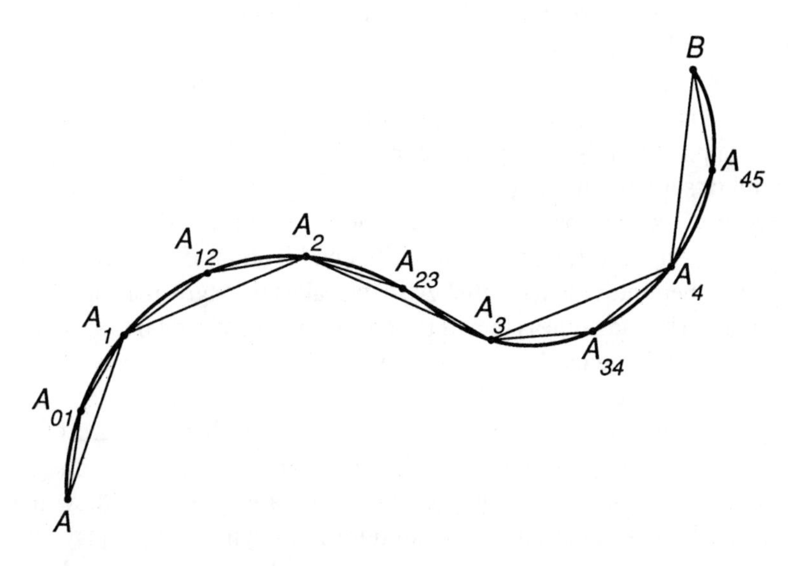

Figure 48 Arc length

roughly follows AB. We can add together the lengths of the straight segments to obtain the length of the broken line $A_0A_1A_2A_3A_4A_5$, and can regard the latter value as an estimate of the quantity we are trying to define. It would surely appear that a better estimate would be gotten by taking more points on the arc, as indicated in Fig. 48, where we have taken an additional point between each adjacent pair of previous points. The new estimate for the length of AB is the sum of the lengths of the straight-line segments A_0A_{01}, $A_{01}A_1$, A_1A_{12}, $A_{12}A_2$, A_2A_{23}, $A_{23}A_3$, A_3A_{34}, $A_{34}A_4$, A_4A_{45}, $A_{45}A_5$. Note that the sum of the lengths of A_0A_{01} and $A_{01}A_1$ is greater than or equal to the length A_0A_1, and likewise for the other pairs into which we have broken the first set of approximating line segments.[1] The process can be continued indefinitely: by taking more and more intermediate points, we get line segments of smaller and smaller size, and the sum of their lengths is greater than or equal to all previous values. But, does this infinite sequence of estimates tend to a definite limit? Not always: there exist curves to which a length cannot be assigned (due to their excessive jaggedness, for example). These are called *non-rectifiable.* For a *rectifiable* curve, however, all approximating sequences of the type discussed above converge to the same limit, and we define this number to be the length l of the curve. In fact, l is the least number that is equal to or greater than the lengths of all approximating broken lines; in other words, it is their least upper bound or supremum (see p. 48).

Once we know how to determine the length of an arc, we can immediately find the length of any finite combination of arcs by adding together their individual lengths. Also, we take the length of an open line segment to be the same as that of the corresponding closed line segment. The length of a general rectifiable curve is the sum of the lengths of its elementary pieces.

It is evident from the discussion in Chapter 7 that each arc of an elementary curve can be deformed into a straight-line segment. Consider the arc AB in Fig. 48 and suppose that it has a length l. We know that AB can be deformed into some straight-line segment $A'B'$ (say). Of course, the length of $A'B'$ is not necessarily l. However, we can always stretch $A'B'$ (or contract it) to become a straight-line segment $A''B''$ of length l. It is therefore clear that every arc can be straightened out to form a straight-line segment having the same length as the arc.[2] This is really why the measuring techniques used in Experiment 20 actually work.

Experiment 21 (Estimating π): Draw a circle as large and as accurately as you can. Inscribe a square in it. Measure the diameter of the circle and the side of the square and evaluate the ratio of the perimeter of the square to the diameter. Next, bisect the sides of the square and extend the bisectors to meet the circumference. Form an inscribed octagon and evaluate the ratio of its perimeter to the diameter. Continue this process for as many stages as you can. What do you find? Try an analytical argument too.[3] □

Formula for Arc Length

When an elementary curve is sufficiently well-behaved, we can obtain a formula for its arc length. To this end, consider an arbitrary point P on an arc AB (Fig. 49) and denote the length of the arc AP by s.

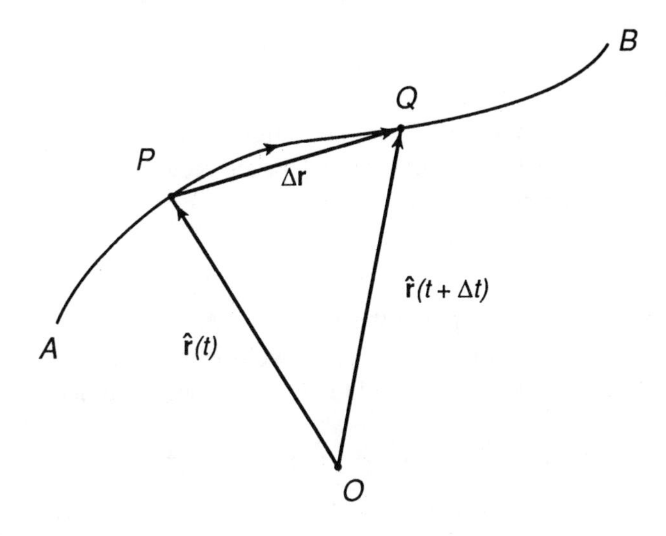

Figure 49 A small increment $\Delta\mathbf{r}$ of arc corresponding to a small increment Δt in the parameter t

Let $\mathbf{r} = \hat{\mathbf{r}}(t)$ be the position vector of P, where t is a parameter used to describe the arc (and can be thought of as time, if you wish), and suppose that t has a value t_1 at A and a value $t_2 > t_1$ at B. The arc is thereby oriented, and we put an arrow on it to signify the direction of traversal. The position vector $\mathbf{r}$ is expressed in component form by Equation (7.2). Corresponding to a small increment Δt ("delta t") in the parameter t, we obtain a point Q not far from P on the curve. Its position vector is $\hat{\mathbf{r}}(t + \Delta t)$. The vector joining P to Q, denoted by $\Delta\mathbf{r}$, is then given by

$$\begin{aligned} \Delta\mathbf{r} &= \hat{\mathbf{r}}\,(t + \Delta\,t) - \hat{\mathbf{r}}(t) \\ &= [\hat{x}\,(t + \Delta t) - \hat{x}(t)]\,\mathbf{i} \\ &\quad + [\hat{y}\,(t + \Delta t) - \hat{y}(t)]\,\mathbf{j} \\ &\quad + [\hat{z}\,(t + \Delta t) - \hat{z}(t)]\,\mathbf{k} \\ &= \Delta x\,\mathbf{i} + \Delta y\,\mathbf{j} + \Delta z\,\mathbf{k}\ , \end{aligned} \tag{8.1}$$

where $\Delta x = \hat{x}(t + \Delta t) - \hat{x}(t)$, *etc.* The length of the chord PQ, which is the same thing as the magnitude $||\Delta\mathbf{r}\,||$ of the vector $\Delta\mathbf{r}$, can be found from the Pythagorean formula

$$||\Delta\mathbf{r}||^2 = (\Delta x)^2 + (\Delta y)^2 + (\Delta z)^2 \ . \tag{8.2}$$

Now, let us denote the length of the arc PQ by Δs. If Q is chosen sufficiently close to P and the curve is of the well-behaved variety, then the length will be approximately equal to the chord length, so that

$$(\Delta s)^2 \approx (\Delta x)^2 + (\Delta y)^2 + (\Delta z)^2 \ . \tag{8.3}$$

Dividing on both sides of the latter expression by the square of the parameter increment Δt, we obtain

$$\left(\frac{\Delta s}{\Delta t}\right)^2 \approx \left(\frac{\Delta x}{\Delta t}\right)^2 + \left(\frac{\Delta y}{\Delta t}\right)^2 + \left(\frac{\Delta z}{\Delta t}\right)^2 \ . \tag{8.4}$$

Now, if we let $\Delta t \to 0$, the quotient $\Delta s / \Delta t$ becomes the derivative ds/dt, and likewise $\Delta x / \Delta t$ becomes dx/dt, *etc.*[4] We then have

$$\left(\frac{ds}{dt}\right)^2 = \left(\frac{dx}{dt}\right)^2 + \left(\frac{dy}{dt}\right)^2 + \left(\frac{dz}{dt}\right)^2 , \qquad (8.5)$$

and hence

$$\frac{ds}{dt} = \sqrt{\left(\frac{dx}{dt}\right)^2 + \left(\frac{dy}{dt}\right)^2 + \left(\frac{dz}{dt}\right)^2} , \qquad (8.6)$$

The relation (8.6) involves derivatives. Formally, we might multiply both sides of this equation by dt to obtain

$$ds^2 = dx^2 + dy^2 + dz^2 . \qquad (8.7)$$

The latter equation is regarded as an expression for the length of an *infinitely small*, or *infinitesimal*, arc. This is a suggestive idea, although it lacks the precision inherent in Equation (8.5).

The length of the arc AB can be written in the integral forms

$$l = \int_{AB} ds = \int_{t_1}^{t_2} \frac{ds}{dt}\, dt . \qquad (8.8)$$

The integral $\int_{AB} ds$ is the limit of the sum of the lengths of inscribed chords which was discussed above. Substituting Equation (8.6) in the first integral in Equation (8.8), we obtain

$$l = \int_{t_1}^{t_2} \sqrt{(\frac{dx}{dt})^2 + (\frac{dy}{dt})^2 + (\frac{dz}{dt})^2}\ dt . \qquad (8.9)$$

Example 1: Arc of a circle. In Example 9 of Chapter 4, we saw that the coordinates of a point on a circle of radius R, centered at the origin, can be expressed in terms of an angle θ $(0 \leq \theta < 2\pi)$ by the equations

$$x = R \cos \theta \quad y = R \sin \theta . \qquad (8.10)$$

Recall from calculus that the derivatives of the sine and cosine functions are the cosine and negative sine functions, respectively. Hence,

$$\frac{dx}{d\theta} = -R \sin \theta \ , \qquad \frac{dy}{d\theta} = R \cos \theta \ . \tag{8.11}$$

With θ as a parameter, Equation (8.6) then becomes

$$\frac{ds}{d\theta} = \sqrt{(-R \sin \theta)^2 + (R \cos \theta)^2} = R \ . \tag{8.12}$$

The length s of an arc subtending an angle between $\theta = 0$ to θ is therefore given by

$$s = \int_0^\theta \frac{ds}{d\theta}\, d\theta = R \int_0^\theta d\theta = R\theta \ . \tag{8.13}$$

Arc Length as a Parameter

As the reader will have observed by now, time is a very useful variable for parametrizing a curve. Angle is also convenient in the case of plane curves. Once the arc length of a curve is available, s may be employed as a parameter. For geometrical discussions, this is often the most suitable choice of parameter. If s is used instead of t, then instead of Equation (7.2), we will have the following description of a curve:

$$\mathbf{r} = \bar{\mathbf{r}}(s) = \bar{x}(s)\,\mathbf{i} + \bar{y}(s)\,\mathbf{j} + \bar{z}(s)\,\mathbf{k} \ . \tag{8.14}$$

In other words, the position vector $\mathbf{r}$ is now given by a function $\bar{\mathbf{r}}$ of arc length, and likewise, the coordinates x, y, z of P are given by functions $\bar{x}, \bar{y}, \bar{z}$ of arc length.

Example 2: A circle parametrized by arc length. From Equations (8.10) and (8.13), we obtain the representation

$$x = R \cos \frac{s}{R} \ , \qquad y = R \sin \frac{s}{R} \tag{8.15}$$

for a circle of radius R centered at the origin. Referring back to Fig. 19, we see that the value $s = 0$ corresponds to the point C, $s = \pi R$ corresponds to B, and $s = 2\pi R$ would give us C again. To occupy all points of the circle once and only once, we may take s to lie in the interval $0 \leq s < 2\pi R$.

Notes to Chapter 8

[1] Recall that any two sides of a triangle are together greater in length than the third side (see *Euclid's Elements*, Book I, Proposition 20 [Heath, 1926]). If A_0A_1 is straight, then equality holds.

[2] Length-preserving deformations are called *isometric* deformations.

[3] By estimating the perimeters of inscribed and circumscribed 96-sided polygons, Archimedes, in his short and brilliant work *Measurement of a Circle*, proved that $3\frac{10}{71} < \pi < 3\frac{1}{7}$. See pp. 91-98 of Heath (1912) and Chap. 4 of Dunham (1990).

9

Tangent

In an important theorem in Book III of his *Elements*, Euclid proves that the straight line drawn at right angles to any diameter *BA* of a circle at one of its extremities *A* falls outside the circle (except where it touches it), and that no other straight line can be interposed into the space *HAE* between the circumference and the straight line, and further that the "horn-like angle" *HAE* that the circumference makes with *EA* is less than any rectilineal angle.[1] The line *EA* is the *tangent* to the circle at *A* (Fig. 50), and, as Euclid's results indicate, it lies closer

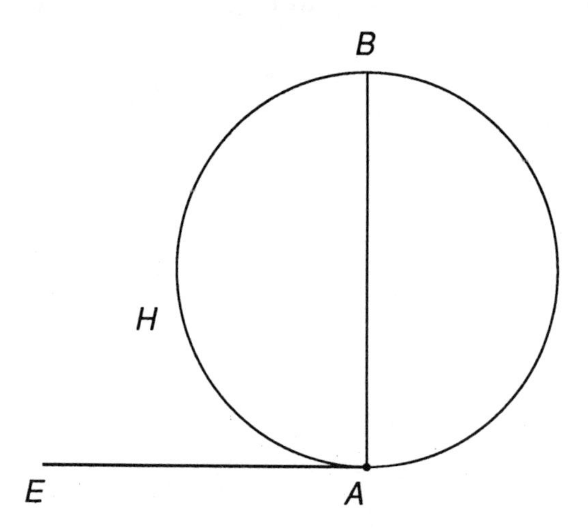

Figure 50 Euclid's construction of a tangent to a circle

to the circumference than any other straight line through *A*. The fact that Euclid felt compelled to *prove* this result, which most of us would regard as "obvious", attests to the high level of rigor that permeated Greek mathematics in the days of Plato's Academy.

The concept of a tangent to a more general curve can be introduced as follows. Consider again an arc AB (Fig. 51) and let P be any point on it except an end-point. Let secants PQ, PR , *etc.*, be drawn through the point P. Notice that as one moves from R to Q and further back along

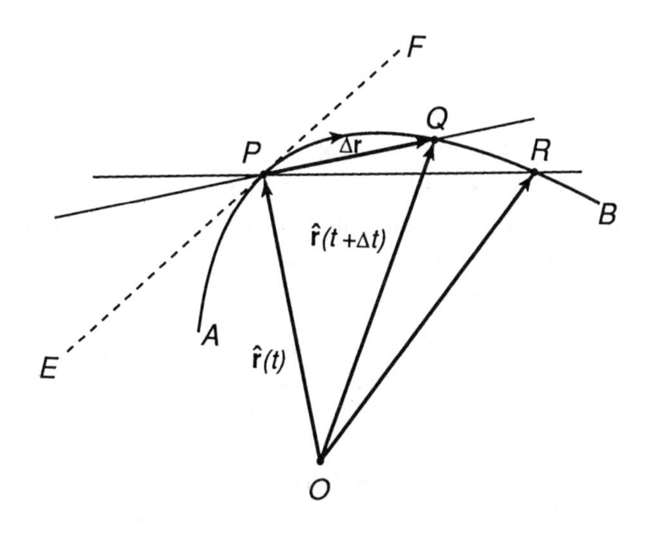

Figure 51 Tangent at a point P on an arc

the arc towards P, the secant rotates. A similar situation occurs if one approaches P from the other end of the arc. For sufficiently nice curves, as one approaches P from either side, the secant gets closer and closer to a definite straight line EF passing through A. We define this limiting line to be the tangent to the arc AB at P.

Experiment 22 (Drawing tangents): By employing the procedure just described, approximately construct the tangent to a circle, and check your accuracy by appealing to Euclid's construction. Try drawing tangents to other curves as well. □

We now wish to express the concept of tangency in algebraic language. Parametrizing the arc by t, we may, as in Equation (8.1), express the vector $\Delta\mathbf{r}$ as

$$\begin{aligned} \Delta\mathbf{r} &= \hat{\mathbf{r}}(t + \Delta t) - \hat{\mathbf{r}}(t) \\ &= \Delta x\, \mathbf{i} + \Delta y\, \mathbf{j} + \Delta z\, \mathbf{k} \ . \end{aligned} \tag{9.1}$$

This represents the difference between the positions P and Q that a moving material point would occupy at times t and $t + \Delta t$, respectively, where Δt signifies a small increment. Note that $\Delta\mathbf{r}$ has the same direction as the secant PQ, and that when we divide $\Delta\mathbf{r}$ by positive increments Δt of the parameter, we still get a vector which is parallel to the secant PQ. Next, take the limit of this quotient as $\Delta t \to 0$ (and hence as Q approaches P), and recall from calculus that this is precisely the derivative $d\mathbf{r}/dx$ of the function $\hat{\mathbf{r}}$:

$$\lim_{\Delta t \to 0} \frac{\Delta\mathbf{r}}{\Delta t} = \frac{d\mathbf{r}}{dt} \ . \tag{9.2}$$

In other words, the derivative of the function $\hat{\mathbf{r}}$, whenever it exists, is a vector parallel to the tangent to the arc AB at P. It is called the *tangent vector*. Relative to the $\{\mathbf{i},\ \mathbf{j},\ \mathbf{k}\}$ basis, we may express it in the form

$$\frac{d\mathbf{r}}{dt} = \frac{dx}{dt}\,\mathbf{i} + \frac{dy}{dt}\,\mathbf{j} + \frac{dz}{dt}\,\mathbf{k} \ . \tag{9.3}$$

When the time parameter t is chosen to parametrize an arc, the tangent vector is identical to the velocity vector: it represents the speed and direction of motion of the moving material point.

Example 1: The straight line. Consider a straight line lying in the x-y plane ($z = 0$). A parametric description of any such line is given by Equations (4.3) and (4.4). The derivatives of x and y are

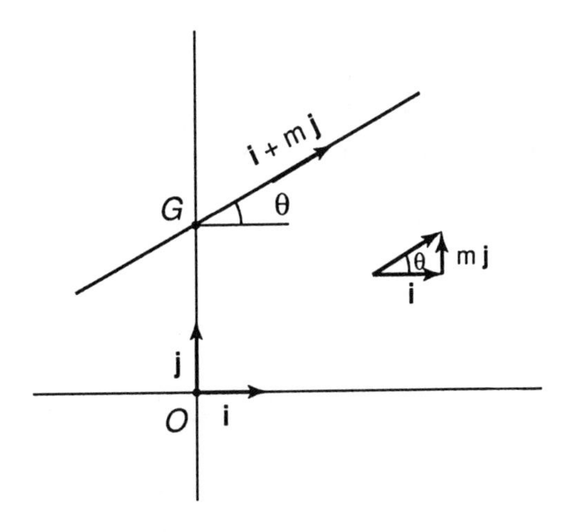

Figure 52 Tangent vector to a straight line

$$\frac{dx}{dt} = 1, \quad \frac{dy}{dt} = m \ , \tag{9.4}$$

m being the slope of the line. At every point on the straight line, the tangent vector is the constant vector

$$\frac{d\mathbf{r}}{dt} = \mathbf{i} + m\,\mathbf{j} \ , \tag{9.5}$$

which is sketched in Fig. 52.

Example 2: The circle. A parametric description of the circle appears in Equations (4.5), the parameter being the angle θ. The position vector of an arbitrary point P on the circle can be expressed as

$$\mathbf{r} = \hat{\mathbf{r}}(\theta) = R(\cos\theta\,\mathbf{i} + sin\,\theta\,\mathbf{j}) \ . \tag{9.6}$$

The tangent vector at P is found by differentiating the function $\hat{\mathbf{r}}$ with respect to θ:

$$\frac{d\mathbf{r}}{d\theta} = R(-\,sin\,\theta\,\mathbf{i} + cos\,\theta\,\mathbf{j}) \ . \tag{9.7}$$

It is sketched in Fig. 53. The vectors $\mathbf{r}$ and $d\mathbf{r}/d\theta$ are perpendicular to one another in this case (observe that their dot product vanishes). Also, note that for each value of θ in the interval $0 \leq \theta < 2\pi$, there is a unique value of the tangent vector $d\mathbf{r}/d\theta$. We therefore have a mapping, which

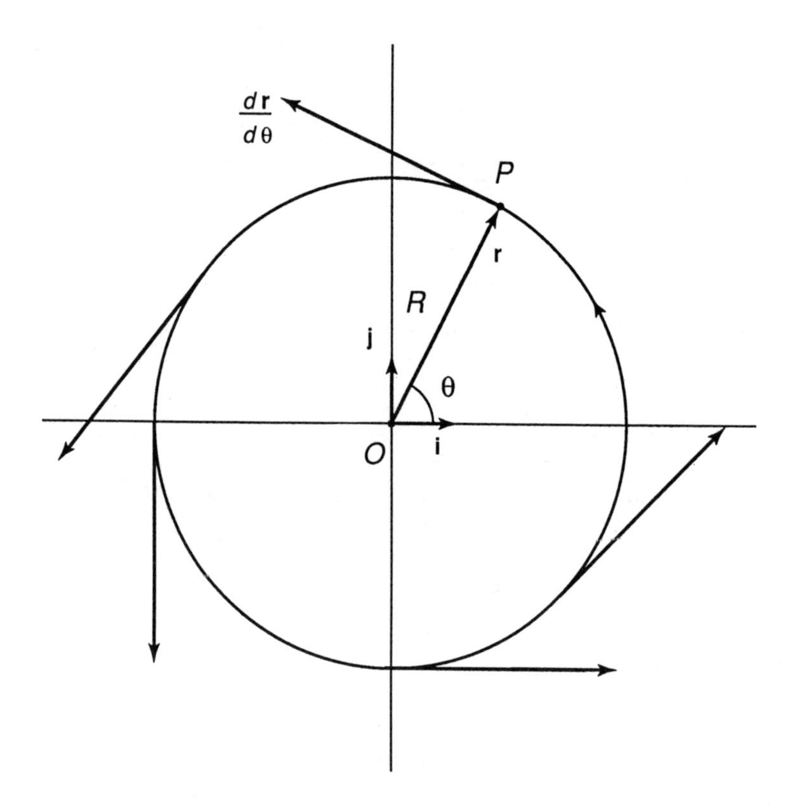

Figure 53 Tangent vector field to a circle

assigns to each θ the corresponding tangent vector. The reader should practice visualizing the entire collection of tangent vectors along a curve. It is customary to refer to this collection as the tangent *vector field.* As θ goes from 0 to 2π radians, the tangent vector rotates continuously through one revolution.

As mentioned at the end of Chapter 8, it is often useful to parametrize curves by their arc lengths. If we use s instead of t in Equation (9.2), we get a tangent vector

$$\frac{d\mathbf{r}}{ds} = \lim_{\Delta s \to 0} \frac{\Delta \mathbf{r}}{\Delta s} . \tag{9.8}$$

Remember, however, that the chord length $||\Delta \mathbf{r}||$ approaches the arc length Δs as Δs tends towards zero (see Equation (8.3)). This forces the tangent vector in Equation (9.8) to be a unit vector. We write

$$\mathbf{t} = \frac{d\mathbf{r}}{ds} = \frac{dx}{ds}\mathbf{i} + \frac{dy}{ds}\mathbf{j} + \frac{dz}{ds}\mathbf{k} , \tag{9.9}$$

and call $\mathbf{t}$ the *unit tangent vector*. In the case of the circle, for example, the parametrization (8.15) leads to the expression

$$\mathbf{t} = \frac{d\mathbf{r}}{ds} = - \sin \theta \, \mathbf{i} + \cos \theta \, \mathbf{j} , \tag{9.10}$$

which is a unit vector lying along the vector $d\mathbf{r}/d\theta$ in Fig. 53.

The tangent vector to a circle rotates in a very agreeable way as we traverse the circumference. What about the tangent vector field for a square? Along each side, between the corners, the unit tangent vector has a constant direction (Fig. 54), but something strange happens at the corners.

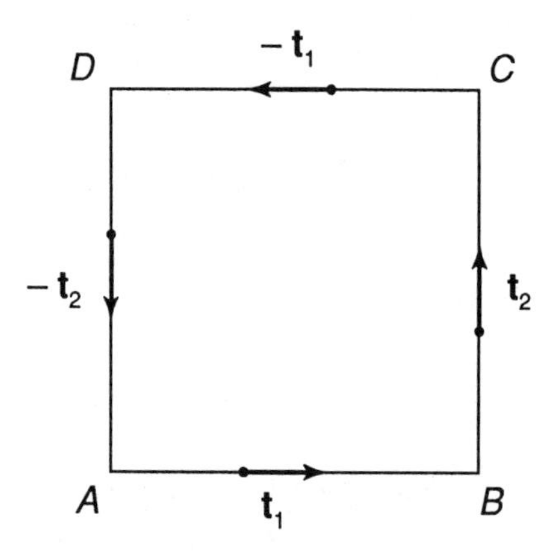

Figure 54 Unit tangent vectors on the sides of a square

It was remarked previously with reference to Fig. 51, that for sufficiently nice arcs, the secant PQ tends to a limiting position EF as Q approaches P from either side. But, we now see that at some peculiar points on a curve the secant PQ may tend to a limiting position EF as P is approached from the left-hand side, whereas a different limiting position $E'F'$ is approached as P is approached from the right-hand side (Fig. 55). It thus seems desirable to enlarge our previous notions of tangency and tangent vector to admit the occasional possibility of having different limiting values as a point is approached from the left and the right. We will call all such points *corners*. If we parametrize the curve in Fig. 55 by its arc length s, reckoned positive in the counterclockwise direction, we will then obtain a unit tangent vector field of the type

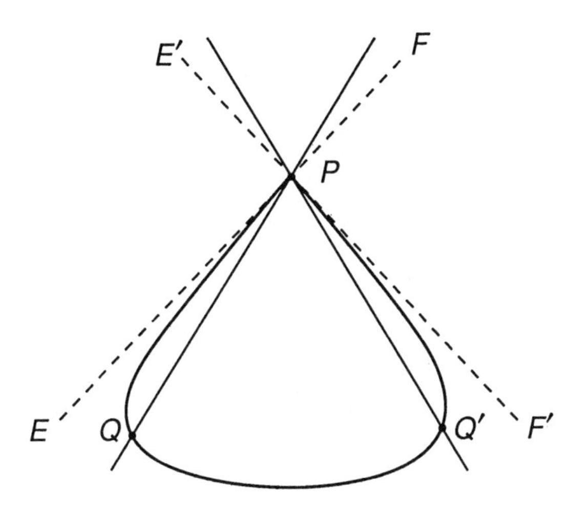

Figure 55 A corner in a plane curve

shown in Fig. 56. The unit tangent vector $\mathbf{t}$ turns continuously as one moves around $OCPBO$, except at the point P, where it turns abruptly, *i.e.,* through a finite angle from its limit $\mathbf{t}_1$ to its limit $\mathbf{t}_2$.

Consider next the curve sketched in Fig. 57, which is obtained from the curve in Fig. 56 by performing a reflection about a horizontal line through P. We may imagine this curve to be traversed from O to C to P, on to M and then R and N, back to P, on to B, and, finally back to O. The unit tangent vector field is sketched in on Fig. 57. Note that if we use the arc length along the path $OCPMRNPBO$ as a parameter, then the first time P is reached, this parameter has a value equal to the length of the arc OCP $(= s_1$, say$)$ and the second time P is reached, the parameter has a value s_2 that is the sum of s_1 and the length of the arc $PMRNP$. Even though the curve has two tangent vectors at the point P, the function now describing it has a unique derivative at s_1 and another unique derivative at s_2.

A situation similar to that of Fig. 57, but now having three unit tangent vectors $\mathbf{t}_1, \mathbf{t}_2, \mathbf{t}_3$ at P, is sketched in Fig. 58.

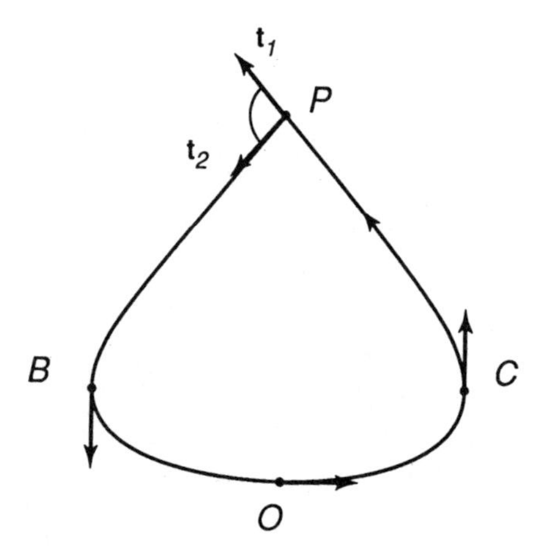

Figure 56 Unit tangent vector field for the curve in Fig. 55

In the case of a cusp, sketched in Fig. 59, we may regard the unit tangent vectors as rotating through 180° at A, since the tangent vector $\mathbf{t}_1$ to the arc BA is horizontal at A and points towards the right, whereas the tangent vector $\mathbf{t}_2$ to AC is also horizontal at A but points towards the left; thus, $\mathbf{t}_2 = -\mathbf{t}_1$. A similar situation occurs when two cusped arcs meet in the manner indicated in Fig. 60.

It is convenient also to use the notion of a one-sided limit to define a tangent vector at the end-points of an arc, such as at A and B in Fig. 51.

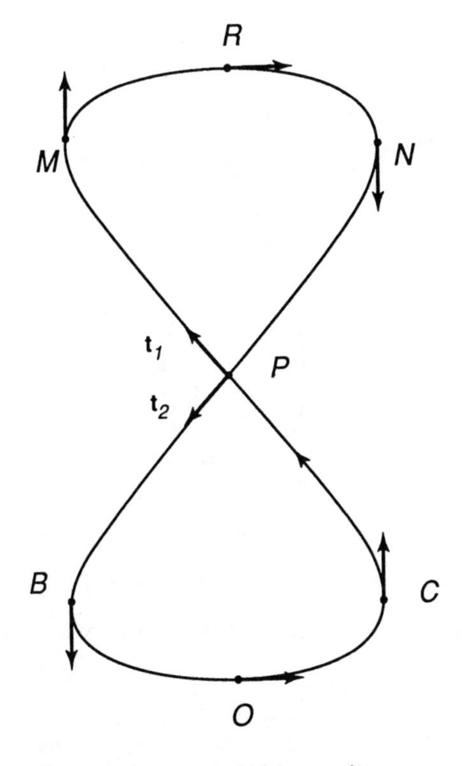

Figure 57 Unit tangent vectors to a self-intersecting curve

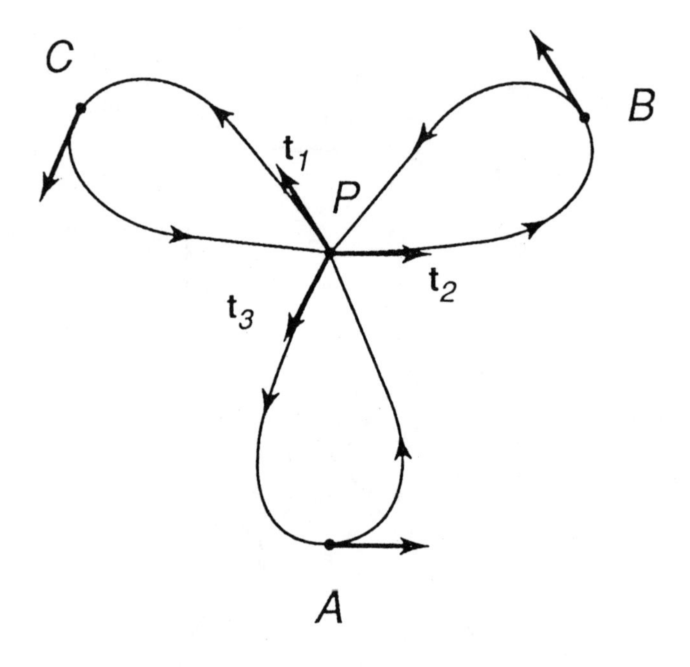

Figure 58 A curve having three unit tangent vectors at P

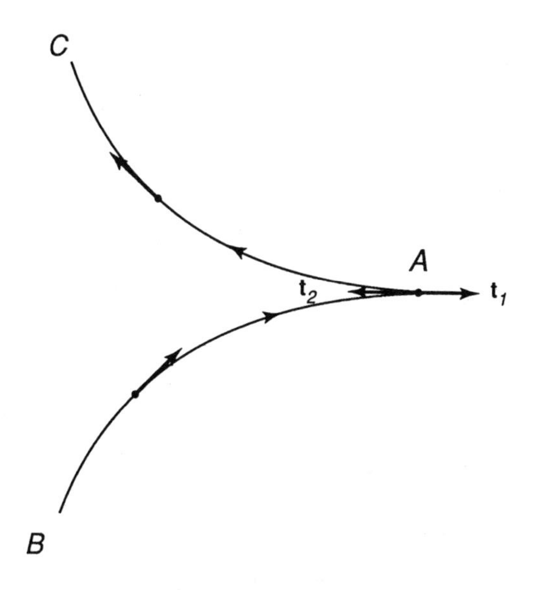

Figure 59 Unit tangent vectors on a curve having a cusp at A

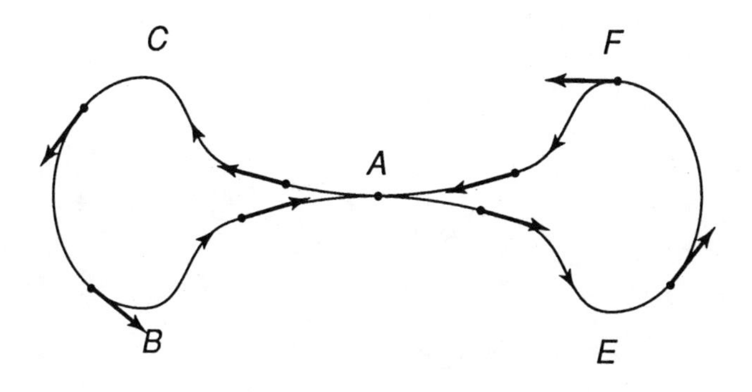

Figure 60 Two curves with cusps, joined to form a curve

Note to Chapter 9

[1] See Proposition 16 and its porism (Book III) in Volume 2 of Heath's 1926 edition of *Euclid's Elements*.

10

Curvature of Curves

> The transition from informal ideas
> to exact measurements and definitions
> represents the transition from a prescientific
> understanding of objects to a scientific theory.
>
> *A.D. Aleksandrov (1963)*

While everyone would agree that a road winding around a mountain is curvy, not quite so apparent is how to express this property quantitatively. Certainly, the curvature of a straight line should be considered zero. Perhaps then, we can regard a curve as a deviation from a straight line? Let us explore how this idea can be given quantitative meaning.

Total Curvature and Average Curvature of a Plane Arc

Let us take a plane arc AB that possesses a unique unit tangent vector at each of its points (Fig. 61). Denote the unit tangent vectors at A and B by $\mathbf{t}_A$ and $\mathbf{t}_B$, respectively, and let $\mathbf{t}$ be the unit tangent vector at an arbitrary point P on the arc. As we move away from the point A towards B, the curve deviates from the tangential direction at A. When B is reached, the tangent to the curve has a direction $\mathbf{t}_B$. We may think of this as the instantaneous direction of the curve itself at B. So, in going from A to B, the curve has changed its direction by an amount equal to the angle through which the tangent vector has rotated. This angle can be easily found from measurements of the angles ϕ_A and ϕ_B that $\mathbf{t}_A$ and $\mathbf{t}_B$ make with an arbitrary chosen fixed direction (Fig. 62). It is also equal to each of the dotted angles at the point where the tangent lines cross one another.

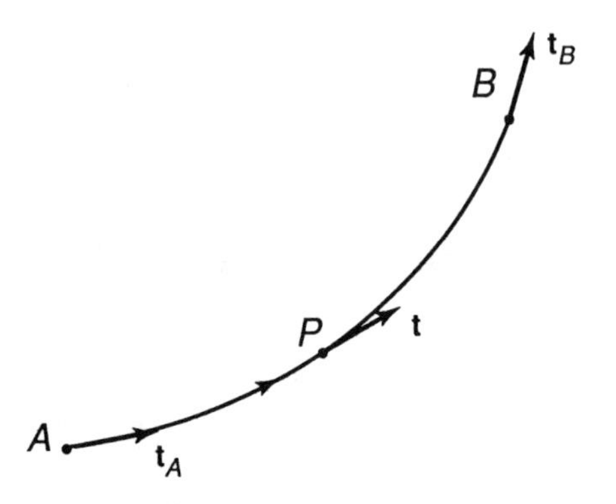

Figure 61 A plane arc and its unit tangent vector field

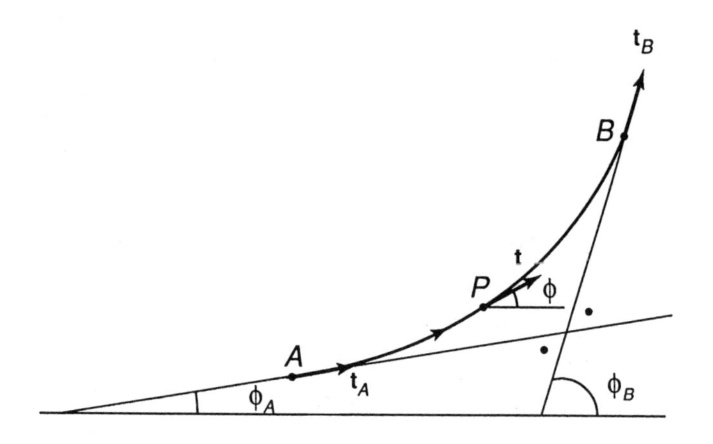

Figure 62 Change in direction of curve

A very useful way of representing the changing direction of a curve is to take an auxiliary unit circle, centered at an arbitrary point, and draw radius vectors in it that are parallel to the tangent vectors to the curve. Since each of these radius vectors has the same magnitude (unity) and direction as the corresponding tangent vector, it is in fact equal to

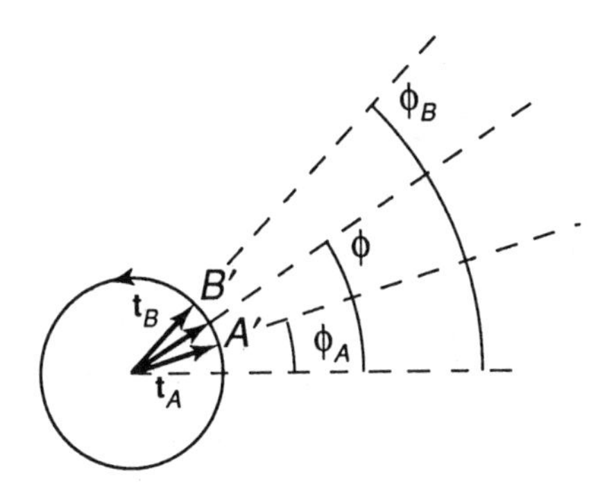

Figure 63 Auxiliary unit circle

that vector. Corresponding to the curve in Fig. 62, we therefore have a diagram of the type shown in Fig. 63. As one moves from A to B along the given curve, the vector **t** in Fig. 63 rotates in the counterclockwise (or positive) direction through an angle $\phi_B - \phi_A$, and the tip of this vector sweeps out an arc $A'B'$ of the unit circle. We define the angle $\phi_B - \phi_A$ to be the *total curvature* of the arc AB:

$$total\ curvature\ of AB = \phi_B - \phi_A . \tag{10.1}$$

Example 1: The straight line. The unit tangent vector to a straight line always maintains the same direction (see Fig. 52), and so the total curvature of every segment of any straight line is indeed zero.

Example 2: Arc of a circle. Consider an arc AB of a circle of radius R (Fig. 64a). Let the angle AOB have measure θ radians. Draw the unit tangent vectors $\mathbf{t}_A$ and $\mathbf{t}_B$. In Fig. 64a, the vector $\mathbf{t}_A$ is vertical. Drawing a vertical line segment BZ through B, it is readily seen that the angle that $\mathbf{t}_B$ makes with BZ also has measure θ. In the auxiliary unit circle in Fig. 64b, copies of $\mathbf{t}_A$ and $\mathbf{t}_B$ are placed, and the angle between them is indicated. The total curvature of the arc AB is θ. If we take the complete circumference, the total curvature is 2π radians.

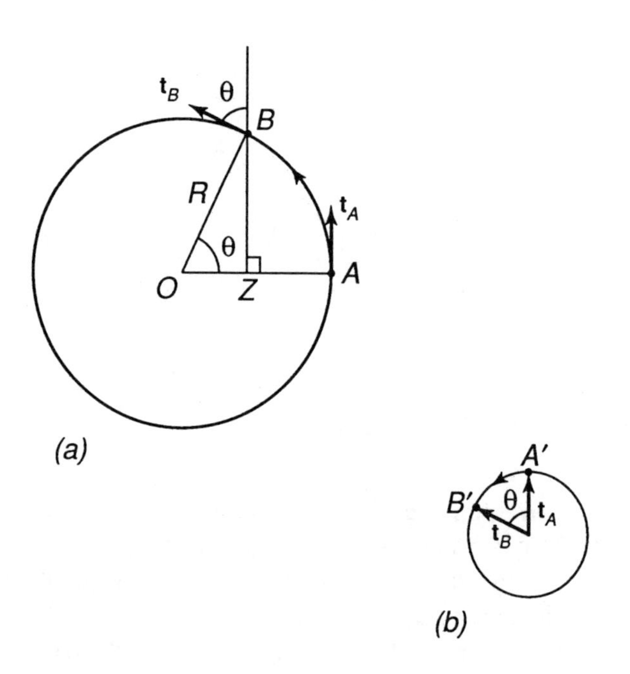

Figure 64 Total curvature of arc of a circle

For the arc sketched in Fig. 62, ϕ_B is greater than ϕ_A, so the total curvature of *AB* is a positive number. For the arc sketched in Fig. 65a, ϕ_B is less than ϕ_A, so the total curvature $\phi_B - \phi_A$ of the arc *AB* is negative in this case. Notice also that the radius vector in the auxiliary unit circle in Fig. 65b now rotates in the clockwise (or negative) direction.

It is obvious that the total curvature of an arc depends only on the direction of the tangent vectors at its end-points. For example, the arc *ACB* in Fig. 66 has the same total curvature as the arc AB in Fig. 62. Also, the two arcs in Fig. 67 with parallel tangents at *B* and *C*, both have total curvatures of $\pi/2$ radians. How can we describe these differences mathematically? One good way of making a distinction between the

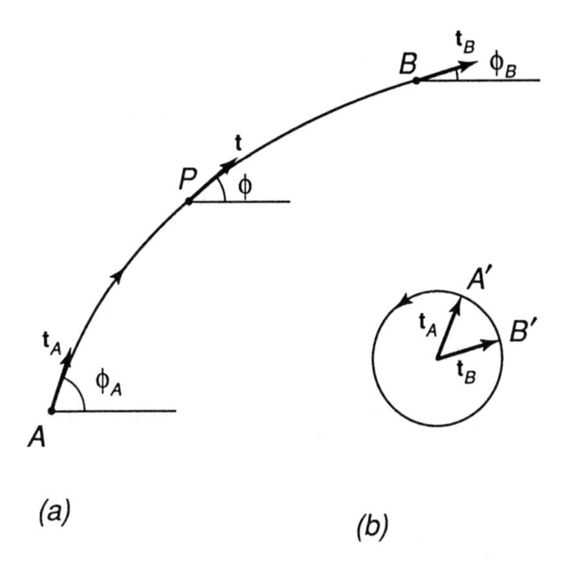

Figure 65 An arc with negative total curvature

"curviness" of the various curves in Figs. 66 and 67 is to compare their total curvatures *over equal lengths* of the curves. Thus, if s_A and s_B are the arc lengths corresponding to two distinct points *A* and *B*, respectively, we define the *average curvature* of the arc *AB* to be

$$\kappa_{avg} = \frac{\phi_B - \phi_A}{s_B - s_A} \quad . \tag{10.2}$$

We note that for an arc of a circle subtending an angle θ, the average curvature is $\theta / R\theta$, or $1/R$.

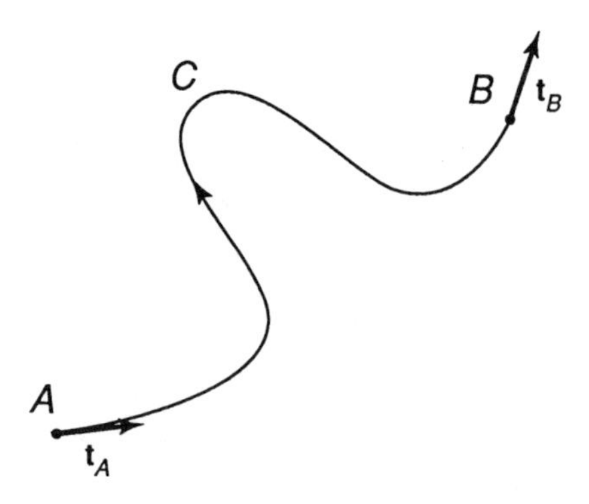

Figure 66 Total curvature of an arc depends only on $\mathbf{t}_A$ and $\mathbf{t}_B$

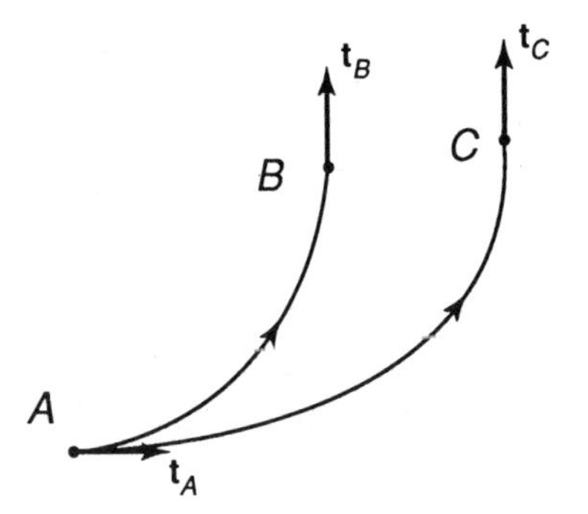

Figure 67 Two arcs having the same total curvature

Experiment 23 (Average curvature of a curve): Draw a curve on a piece of cardboard. Cut along the curve and lightly sand the edge. Trace the curve on a sheet of paper. Lay a thread alongside the cardboard to measure lengths of arc to various points on the curve. Draw tangents at these points and measure the angles they make with a fixed direction. Estimate the average curvature of a few segments of the curve. □

The reader will now appreciate how the formula (10.2) enables us to attain a quantitative understanding of the manner in which a curve curves. He or she may also have wondered about what happens to κ_{avg} as smaller and smaller segments of a curve are taken. This is where calculus again comes to the rescue. Consider any point P on an arc AB of a curve (Fig. 68a) which has been parametrized by its arc length, the value of which is s at P. The unit tangent vector at P is $\mathbf{t}(s)$, and it makes an angle $\phi(s)$ with the horizontal direction. Take an increment Δs of arc length such that Q is a point of the arc AB and may lie on either side of P. The unit tangent vector at Q is $\mathbf{t}(s + \Delta s)$ and it is inclined

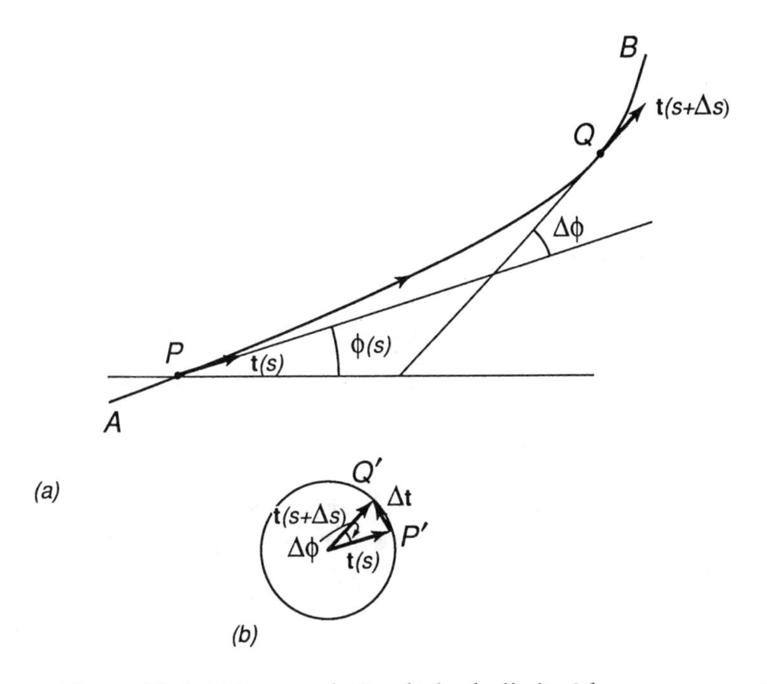

Figure 68 Let Q approach P and take the limit of the average curvature of PQ.

at an angle $\phi(s + \Delta s)$ to the horizontal direction. The angle through which the tangent rotates as one moves along the curve from P to Q is $\phi(s+\Delta s) - \phi(s)$, which we shall write as $\Delta\phi$ (see Fig. 68a). The tangent vectors $\mathbf{t}(s)$ and $\mathbf{t}(s + \Delta s)$, and the angle $\Delta\phi$, are shown again in the auxiliary unit circle in Fig. 68b. In accordance with Equation (10.2), the average curvature of the arc PQ is $\Delta\phi/\Delta s$. Taking a hint from calculus, we let Q lie closer and closer to P. Then, if the quotient $\Delta\phi\ /\Delta s$ tends to a limiting value κ, we define this to be the *curvature of the curve at P*. Thus, we have

$$\kappa = \lim_{\Delta s \to 0} \frac{\Delta\phi}{\Delta s} = \frac{d\phi}{ds}\ , \qquad (10.3)$$

i.e., κ is the derivative of the angle ϕ, regarded as a function of arc length. For example, in the case of a circle of radius R, since $\Delta s = R\ \Delta\phi$, we have

$$\begin{aligned} \kappa &= \lim_{\Delta s \to 0} \frac{\Delta s\ /R}{\Delta s} = \frac{1}{R} \lim_{\Delta s \to 0} \frac{\Delta s}{\Delta s} \\ &= \frac{1}{R}\ . \end{aligned} \qquad (10.4)$$

In other words, at every point on a circle, the curvature is $\frac{1}{R}$. Smaller circles have greater curvature than larger ones. For every point on a straight line, $\kappa = 0$ (since $\kappa_{avg} = 0$). In general, κ can take on positive, negative, and zero values along an arc. A point at which κ vanishes is called a *point of inflection*. The arc in Fig. 69, for instance, has negative curvature at points between A and C, positive curvature between C and B, and there is a point of inflection at C.

The definition in Equation (10.3) is indeed a beautiful one: It captures a sound physical intuition in a precise mathematical form. Once again, one is reminded of Lebesgue's observation (see p. 92) about geometrical measurements beginning physically, but ending metaphysically.

The formula (10.4) suggests the following idea: at any point P on an arc at which the curvature is positive, we could place for comparison a circle of radius $1/\kappa$; this circle would have the same curvature as the

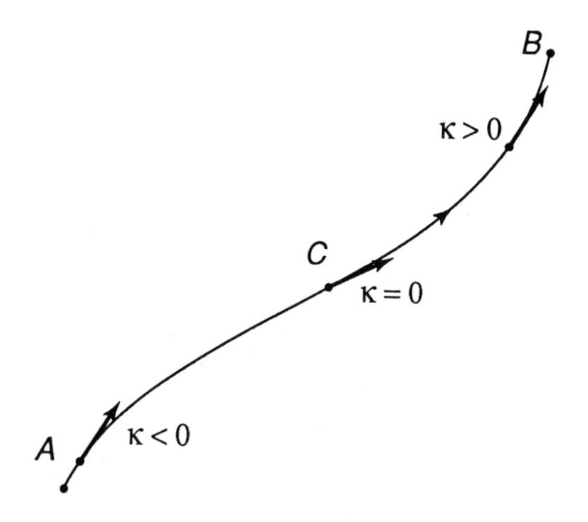

Figure 69 Variation of curvature along an arc

arc has at P. Let us construct it in such a way that at P it shares a tangent with the arc, and that its center lies on the side towards which the arc is turning (*i.e.*, on the left for $\kappa > 0$). Thus, in Fig. 70, we erect a perpendicular to the tangent at P and find the center of the circle at C, where CP has length $1/\kappa$. At any point P' at which the curvature κ' of the arc is negative, we may take the absolute value $|\kappa'|$ of κ' and set $R' = 1/|\kappa'|$; the center C' of the corresponding circle is taken to lie on the right side of the arc. The circle constructed in the foregoing manner is called the *circle of curvature*, and its radius is called the *radius of curvature* (at P).

Another way of thinking about the circle of curvature is this: Let Q and R be two other points on the arc, as shown in Fig. 71, and construct a circle that passes through P, Q, R; its center lies at the intersection of the perpendicular bisectors of QP and PR. Now, let Q and R approach P simultaneously, but independently of one another. The limiting position of this sequence of circles is the circle of curvature at P.

Experiment 24 (Circle of curvature): For the curve you used in Experiment 23, construct a series of circles, each of the type shown in

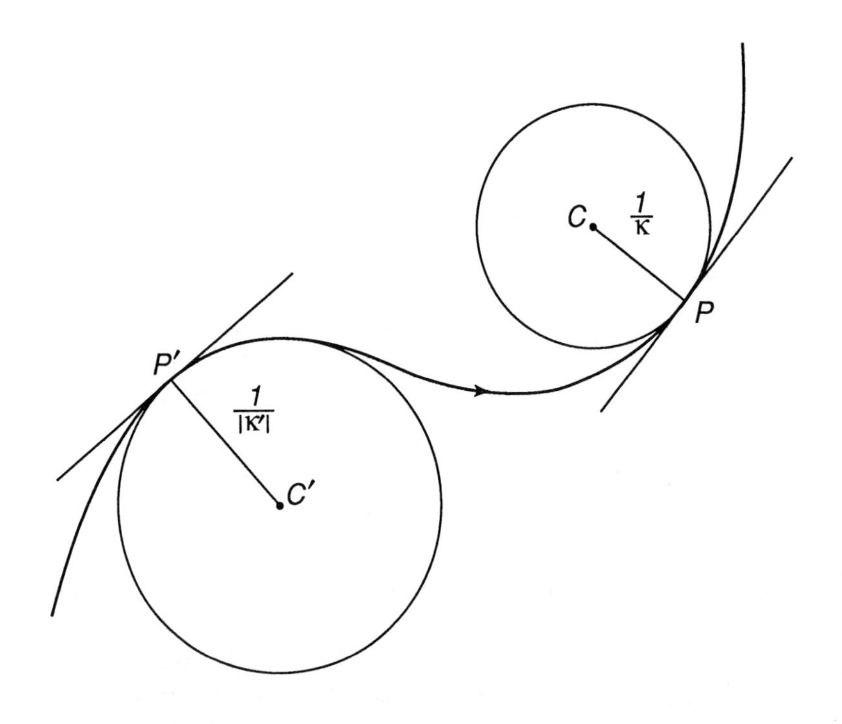

Figure 70 Circles of curvature at P and P'

Fig. 71, but on diminishing arcs QR. In this manner, approximate the circle of curvature at a point P. Compare the reciprocal of the radius of each of your circles to the average curvature of the arc upon which it stands.□

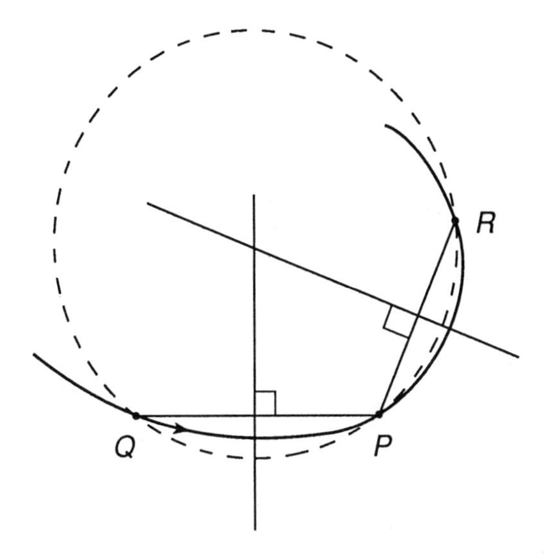

Figure 71 A circle passing through P, Q, and R

Rotation Index of a Plane Closed Curve

We now come to a very interesting idea. Suppose that we follow the unit tangent vector $\mathbf{t}$ all the way around a plane closed curve, such as the curve *ABCDEFGHA* in Fig. 72. At A, $\mathbf{t}$ is horizontal; its value is $\mathbf{t}_A$, which we also represent on the auxiliary unit circle in Fig. 72. As we move towards B, $\mathbf{t}$ rotates counterclockwise through 90°. Between B and C, it first rotates counterclockwise, but then clockwise until it is vertical again, so that $\mathbf{t}_C = \mathbf{t}_B$. Continuing on around the curve, we see that $\mathbf{t}$ performs a net counterclockwise rotation of 360°, or one complete revolution, around the unit circle. We say that the *rotation index* of $ABCDEFGHA$ is 1.

Consider next the curve shown in Fig. 73, which has loops at B, C, D, E. If we follow the tangent vector $\mathbf{t}$ around this curve starting from A, we see that $\mathbf{t}$ certainly returns to its horizontal position at A. However,

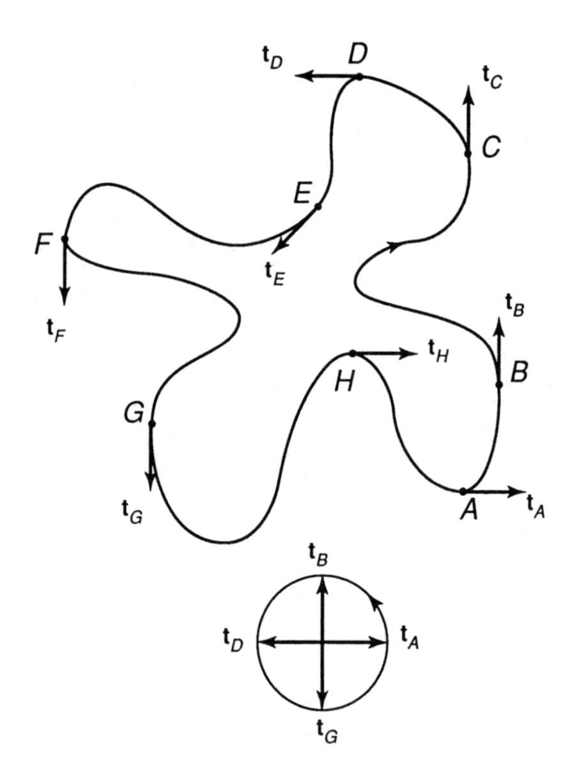

Figure 72 Following the unit tangent vector around a closed curve

as we travel around the loop at B, $\mathbf{t}$ performs a full counterclockwise revolution. Likewise, on each of the loops at C, D, and E, another full counterclockwise rotation occurs. Upon its return to A then, five full rotations of the tangent vector have taken place: the rotation index of $ABCDEA$ equals 5. Another example is shown in Fig. 74: here, the tangent vector $\mathbf{t}$ performs 1 counterclockwise revolution on the upper loop $ABCDA$, but it rotates through 1 clockwise revolution on the lower loop $AEFGA$ (which we count as a -1 counterclockwise revolution); the rotation index of the figure eight is therefore 0.

Experiment 25 (Rotation index): Create examples of the type appearing in Figs. 72, 73, 74 to yield rotation indexes of 2, 3, and -1. □

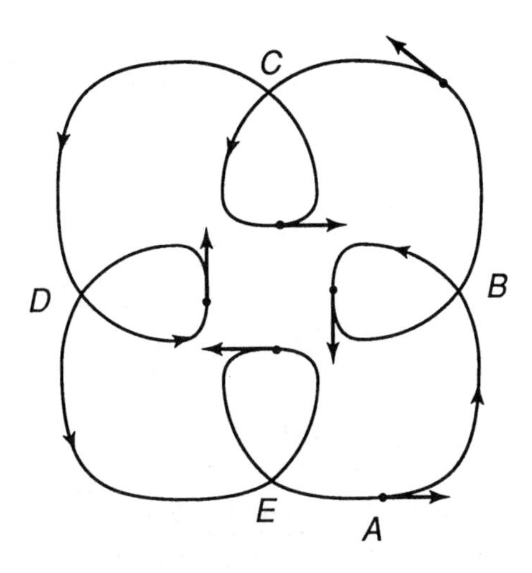

Figure 73 Rotation index of a curve with loops

The examples we have been considering all suggest that the total curvature of a closed plane curve is always some integer times 2π radians, the integer being the rotation index, *i.e.*,

$$rotation\ index = \frac{total\ curvature\ of\ closed\ curve}{2\pi}. \qquad (10.5)$$

Towards the end of Chapter 9, we discussed curves at certain points of which the tangent vector abruptly changes its direction. Thus, for the square in Fig. 54, **t** rotates counterclockwise through $\pi/2$ radians at each of the corners A, B, C, D. Since **t** does not rotate at all in between consecutive corners, we see that its net rotation as we completely traverse the square is 1 counterclockwise revolution. Thus, its rotation index is 1, which is the same as for a circle (Fig. 64) and for the curve *ABCDEFGHA*

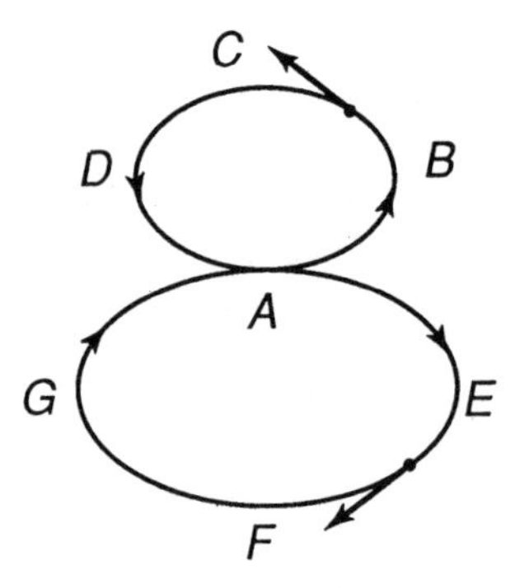

Figure 74 Rotation index of a figure eight

in Fig. 72. Similarly, each of the two curves in Fig. 75 has a rotation index equal to 1. In fact, any Jordan curve in the plane has a rotation index equal to 1, when traversed in the counterclockwise direction. Hence, the rotation index is 1 for any plane curve that can be obtained by deforming a circle. This is quite a remarkable result.

The fact, that for a triangle (Fig. 76) the rotation index is 1, can be exploited to furnish a nice proof of the theorem on angle sum of a triangle (see p.12). Thus, starting out at A, we follow the unit tangent vector $\mathbf{t}$ along the side AB, where it experiences no rotation. At B, it rotates clockwise through an angle θ_2 (radians) to line up with the side BC. Similarly, its experiences no rotation along BC, but swings through an angle θ_3 at C. Finally, along CB it does not rotate, but rotates through θ_1 at A. Its original portion is now restored, and we therefore have

$$\theta_1 + \theta_2 + \theta_3 = 2\pi \,. \tag{10.6}$$

But,

$$\theta_1 + \alpha = \pi \,, \; \theta_2 + \beta = \pi, \quad \theta_3 + \gamma = \pi \,. \tag{10.7}$$

Consequently,

$$\alpha + \beta + \gamma = \pi. \tag{10.8}$$

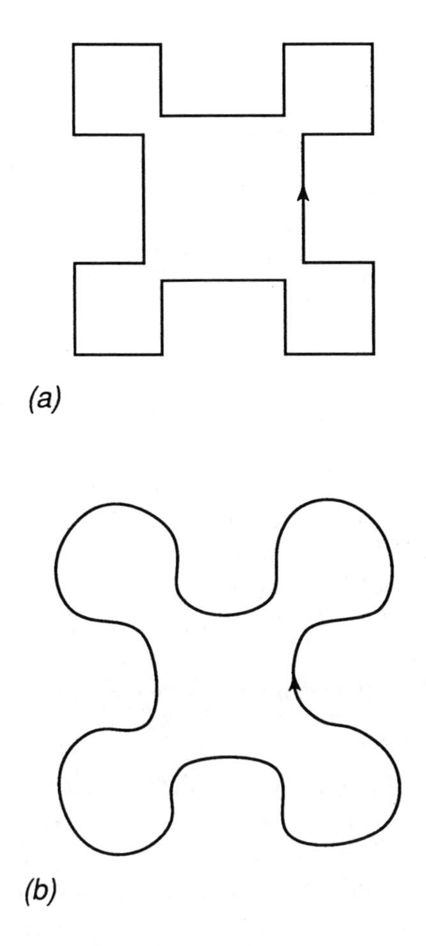

Figure 75 Two curves having rotation index 1

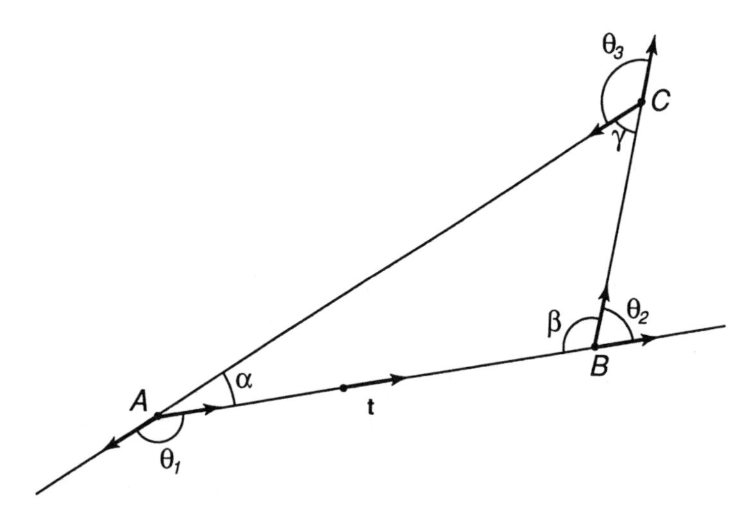

Figure 76 Following the unit tangent vector **t** around a triangle

Experiment 26 (Angle sum for a convex polygon): (a) Draw a convex polygon having 7 sides. (Convexity can be taken to mean that each interior angle is less than π radians in measure.) Measure the interior angles and sum them.

(b) Use the fact that the rotation index of the polygon is 1 to establish the exact answer.

(c) Use the same argument to find the angle sum for a convex polygon with n sides. □

It was mentioned above that the rotation index of every plane curve that is obtainable by deforming a circle is equal to 1 (when traversed in the counterclockwise direction). It is very interesting to examine how differently the curvature κ may be distributed in different members of this family of curves, even though the total curvature for each one of them is 2π. To elaborate, we note that for a circle the curvature has the same constant value – given by Equation (10.4) – at all points on its circumference: its total curvature is evenly distributed over the circumference. For the curve in Fig. 75b, which can be obtained by deforming a circle, the curvature varies as one proceeds around the curve.

For the rectilinear figure in Fig. 75a, κ is zero in between consecutive corners. At the corners, κ is not defined at all, as can be seen from an examination of Equation (10.3): take a small segment of the perimeter containing a corner; let its length be Δs, and suppose that the unit tangent vector rotates through $\pi/2$ radians at the corner. Then, if κ existed at the corner, it would be equal to the limit of $2\pi/\Delta s$ as Δs tends to zero. But $2\pi/\Delta s$ tends to infinity (which is not a number) as Δs tends to zero. Hence, there is no real number that satisfies Equation (10.3) at the corner. We may say that there is a concentration of total curvature at each corner. Since κ may be regarded as a density of total curvature (*i.e.*, total curvature per unit arc length), it makes sense that κ should tend to infinity at corners.

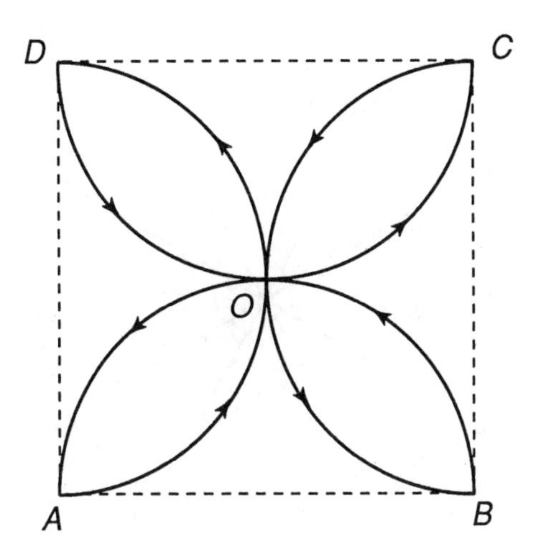

Figure 77 A curve formed from semicircles built on the sides of a square

As you slowly deform a circle into a triangle (say), you are causing a redistribution of curvature to take place. At the outset, κ is the same everywhere on the circle, then differences in curvature gradually arise between different points. As the circle approaches a triangular share, κ approaches zero at points which will lie on the sides of the triangle, and κ becomes very large at the three points that will become vertices of the triangle. By straightening out arcs of the circle to become sides of a triangle, you drive the total curvature of the arcs into the vertices of the triangle.

Experiment 27 (Rotation index for a family of curves): (a) On the sides of a square construct semicircles, as shown in Fig. 77. Calculate the rotation index for the closed curve $OAOBOCODO$, when traversed in the counterclockwise direction. Explain how total curvature is distributed over this curve.

(b) Argue that the curve $OAOBOCODO$ in Fig. 77 can be deformed

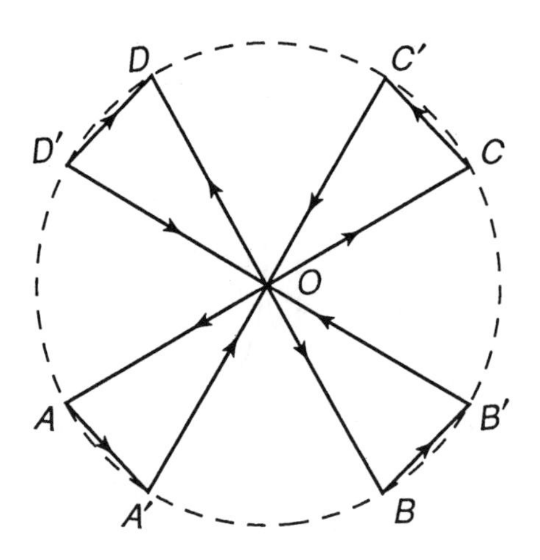

Figure 78 A close relative of the curve in Fig. 77

into the rectilinear figure $OAA'OBB'OCC'ODD'O$ in Fig. 78. The latter figure consists of four congruent isosceles triangles spaced evenly around a circle; the angle AOA' has measure $30°$. Calculate the rotation index of $OAA'OBB'OCC'ODD'O$. What do you conclude? Describe how total curvature is distributed along $OAA'OBB'OCC'ODD'O$.

(c) Using small nails, a board, and a large rubber band, make a model of the curve in Fig. 76. Stretch the rubber band to generate other members of this family of curves. □

Additional Remarks on Curvature of Curves

Remark 1. Referring to Fig. 68b, we note that $\Delta\phi$ (in radian measure) has the same value as the length of the arc $P'Q'$ of the auxiliary unit circle. We are therefore led to another important interpretation of the average curvature of the arc PQ: it is the quotient of the length of the arc $P'Q'$ on the auxiliary circle and the length of the arc PQ on the curve whose curvature is being investigated. Further, the curvature κ is the limit of the quotient of these two lengths.

Remark 2 . The total curvature of an arc AB can be written in terms of the curvature κ as an integral, namely

$$\phi_B - \phi_A = \int_{s_A}^{s_B} \frac{d\phi}{ds}\, ds = \int_{s_A}^{s_B} \kappa\, ds \quad , \tag{10.9}$$

where s_A and s_B are the values of the arc lengths at A and B. For this reason, the total curvature is also referred to as the *integral curvature* of an arc.

Remark 3. Suppose that we take the derivative of the unit tangent vector $\mathbf{t}(s)$. This is given by the expression

$$\frac{d\mathbf{t}}{ds} = \lim_{\Delta s \to 0} \frac{\mathbf{t}(s + \Delta s) - \mathbf{t}(s)}{\Delta s} \,. \tag{10.10}$$

The quantity $\mathbf{t}(s + \Delta s) - \mathbf{t}(s)$ is the change in the unit tangent vector as one goes from P to Q and is represented by the vector $\Delta\mathbf{t}$ joining P' to Q' on the auxiliary circle in Fig. 68b. The derivative $d\mathbf{t}/ds$ is therefore a measure of the rate at which the unit tangent vector changes its direction as one moves along the curve. There must be a very close connection between this derivative and the curvature κ, defined by Equation (10.3). To see what the connection is, consider any point that is not a point of inflection (*i.e.*, $\kappa \neq 0$), and recall that the length of the arc $P'Q'$ in Fig. 68b is then $\Delta\phi \neq 0$. Therefore, the magnitude of $\Delta\mathbf{t}$ divided by $\Delta\phi$ will tend to 1 as $\Delta s \to 0$, and $\Delta\mathbf{t}/\Delta\phi$ will tend to a unit vector, which we will call $\mathbf{n}$. The direction of $\mathbf{n}$ is tangent to the unit circle at P' in Fig. 68b. Rewriting Equation (10.10) in the form

$$\frac{d\mathbf{t}}{ds} = \lim_{\Delta s \to 0} \left(\frac{\Delta\mathbf{t}}{\Delta\phi}\frac{\Delta\phi}{\Delta s}\right), \tag{10.11}$$

and making use of Equation (10.3), we will arrive at the expression

$$\frac{d\mathbf{t}}{ds} = \kappa\,\mathbf{n} . \tag{10.12}$$

We may now think of $\mathbf{n}$ as a unit vector which is perpendicular, or *normal,* to the tangent to the curve AB at P in Fig. 68a. It is called the *principal unit normal vector*. The vector $d\mathbf{t}/ds$ is called the *curvature vector* (and is often denoted by a separate symbol). Equation (10.12) has been derived on the assumption that $\kappa \neq 0$: in this case, the unit vector $\mathbf{n}$ is uniquely determined by $d\mathbf{t}/ds$ and κ . (At points of inflection, both $d\mathbf{t}/ds$ and κ vanish, and Equation (10.8) will hold no matter what unit vector is substituted for $\mathbf{n}$.)

Remark 4. Referring to Fig. 79, we can now associate a pair of mutually perpendicular unit vectors $\mathbf{t}$ and $\mathbf{n}$ with each point on an arc. As we move along the arc, the pair rotates. The motion of the two vectors can be represented on an auxiliary circle, as also indicated in the figure. Note that the arc traced out on the unit circle by the vector $\mathbf{n}$ will have the same length as that traced out by $\mathbf{t}$.

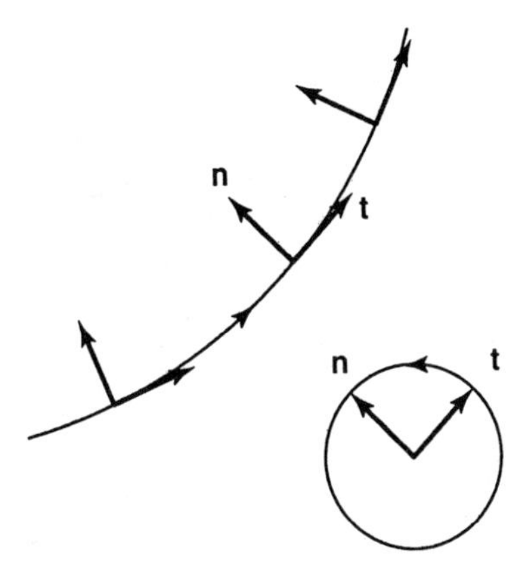

Figure 79 Unit tangent and principal unit normal vector fields

Space Curves

In our discussion so far, the concept of curvature has been treated only in the context of plane curves. Can we extend this treatment to space curves, such as the arc sketched in Fig. 80? Yes, we can, but some modifications have to be made in our previous development.

The approach is conceptually the same, however: we watch how the tangent turns as we move along the curve. Now, the definition given in Equation (9.8) for the unit tangent vector **t** already applies to a space curve, so no change is needed there. Tracking the direction of **t** can no longer be done by means of a single angle ϕ as we had in Fig. 62, because **t** is no longer confined to move in a fixed plane. We could use two angles to keep track of its position.[1] However, it is better to deal directly with the derivative $d\mathbf{t}/ds$, which captures how **t** changes as one moves along the curve. The definition which we gave for this quantity in Equation (10.10), when we were thinking only about plane curves, actually applies just as well to space curves. So, $d\mathbf{t}/ds$ is the curvature vector for a space curve, too. The derivation leading up to Equation (10.11) no longer holds, but it so happens that even for a space curve,

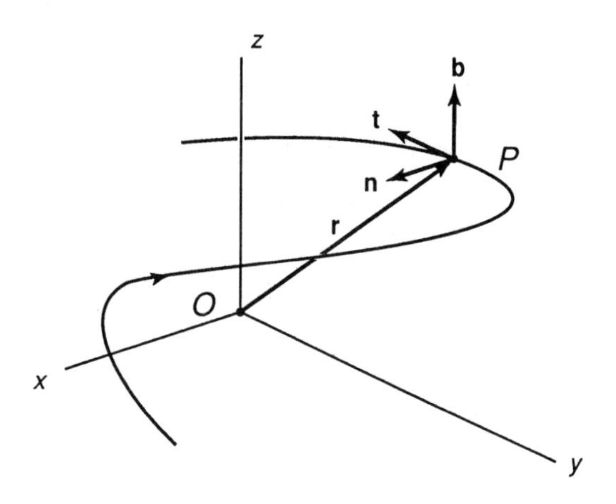

Figure 80 A space curve and its Serret-Frenet triad

the curvature vector can still be expressed in the same form as Equation (10.12), *i.e.*, as a real number κ times a unit vector **n**. At each point of the arc in Fig. 80, we will have a unit tangent vector **t** and a principal unit normal vector **n**. The plane of **t** and **n**, called the *osculating plane*, is no longer a fixed plane, but instead, rotates as we move along the curve. Let **b** be a unit vector perpendicular to the osculating plane, drawn as shown in Fig. 80. This vector is called the *unit binormal vector*. So, at each point along the curve, we now have a triad of unit vectors, **t**, **n**, **b**, making right-angles with one another.

The imaginative reader will surely ask himself or herself: "How do the vectors **t**, **n**, **b** change as I go along the curve?" This is a splendid question, and leads to a magnificent set of equations, which were derived independently by the French geometers J.-F. Frenet and J.A. Serret in the middle of the 19th century.[2] Although we shall not give a derivation of these equations in the present book, it would be a pity not to write them down, especially since it should now be possible for the reader to appreciate what these formulae are saying. The three *Serret-Frenet formulae* for space curves are:[3]

$$\begin{aligned}
\frac{d\mathbf{t}}{ds} &= \kappa\,\mathbf{n} \quad , \\
\frac{d\mathbf{n}}{ds} &= -\kappa\,\mathbf{t} + \tau\,\mathbf{b}\,, \\
\frac{d\mathbf{b}}{ds} &= -\tau\,\mathbf{n}\ .
\end{aligned} \tag{10.13}$$

The first of these we have met before as Equation (10.12): it states that the rate at which **t** is changing with respect to arc length s is equal to the curvature κ times the principal unit normal **n**. The last formula of the three shows that the rate at which the binormal **b** is changing can be expressed as a real number τ, called the *torsion*, times the unit vector $-$ **n**. Since **b** is perpendicular to the osculating plane, this equation implies that for $\tau \neq 0$ the normal to the osculating plane turns in the direction opposite to the unit vector **n**. For a plane curve, both $d\mathbf{b}/ds$ and the torsion vanish. The second formula in the set describes how the principal normal is changing as a function of arc length: it has two components–one points opposite to **t**, and the other points along **n**. The Serret-Frenet formulae are the main tool for the analytical study of the geometry of curves, and are extremely useful also in studying the motion of particles in dynamics, and in treating rod-like components of engineering structures and machines.

Notes to Chapter 10

[1] In the planar case, the tip of the unit tangent vector **t** lies on the auxiliary unit circle; in the case of space curves, the tip of **t** lies on an auxiliary unit sphere.

[2] J.-F. Frenet (1816-1900) presented a set of formulae for the geometry of space curves as part of his doctoral dissertation at the University of Toulouse in 1847, and in 1852 published these results in the *Journal de Mathématiques Pures et Appliquées*. In the meantime, J.A. Serret (1819-1885), a mathematician in Paris, published his formulae in the same journal in 1851.

[3] For a derivation of these formulae, see Struik (1961), Lipschutz (1969), or O'Neill (1966).

11

Surfaces

Having worked his or her way through Chapters 7 to 10, the reader will have gained an appreciation for the geometry of curves. We now turn our attention to the geometry of surfaces, which is even more fascinating. We will proceed in our usual manner, moving from intuitions to concepts, and exploring the geometrical phenomena by means of simple experiments. Our discussion of surface geometry begins with a search for a good definition of the concept of a surface.

What is a Surface?

As in the case of curves, we all "know" what a surface is, but it requires some effort to make this intuitive idea precise. We will build up a satisfactory definition gradually.

Consider a square $A'B'C'D'$ (Fig. 81), made of highly elastic material, such as you used in Experiment 5. Let it be moved through space, bent, stretched or contracted, in any way that does not involve overlapping, tearing, or puncturing. Such a process, as we know by now, is a *deformation.*[1] A configuration resulting from a deformation of $A'B'C'D'$ is given by *ABCD*. Intuitively, one would certainly want to call *ABCD* a surface. Straight lines inscribed on $A'B'C'D'$ have become curves , and the right-angled triangle $P'Q'R'$ has been distorted into the curvilinear three-sided figure *PQR*. The angle between the surface curves *PQ* and *PR* in Fig. 81 may be regarded as the angle which the tangents to these curves at *P* make with one another.

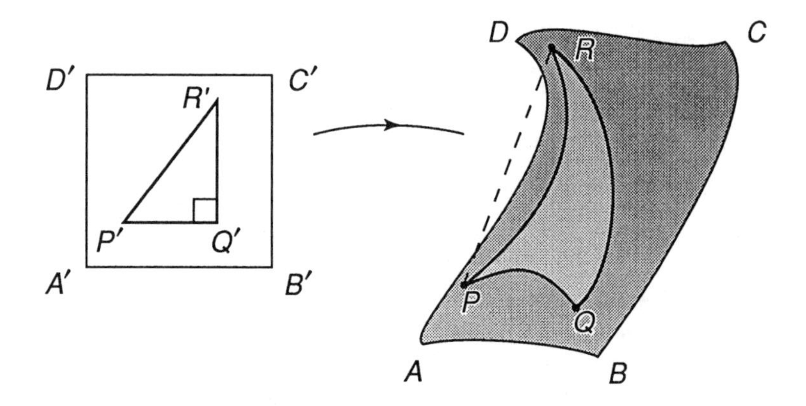

Figure 81 The square $A'B'C'D'$ is deformed into a surface $ABCD$

To measure the length of a curve *PR* on the surface $ABCD$, lay an inextensible thread along the curve, mark it off at *P* and *R*, remove it from the surface, straighten it, and measure the distance between the marks with a ruler. Except in special cases, the length of a curve that lies in the surface and joins *P* to *R* is not equal to the length of the dashed straight line joining *P* to *R* in Fig. 81. This independence of surface measurements from space measurements carries with it the potential for a surface geometry that is essentially different from the Euclidean geometry of the three-dimensional space in which the surface is located.

In Fig. 82, a deformation is indicated which takes a square $A'B'C'D'$ with a hole H' into a deformed configuration *AEBFCGIDJ* with a hole *H*. A sharp bend has appeared along *JF*, and new corners have been formed at *E, F, G,I,J*. However, no puncturing, tearing, or overlapping has occurred. Despite the bend and the corners, one would like to regard *AEBFCGIDJ* as a surface.

Scrutinizing Fig. 82 further, we observe that if *P* is any point in the deformed configuration $AEBFCGIDJ$, then there is a unique point P' in $A'B'C'D'$ which has been mapped into *P*; moreover, for any piece $\mathcal{N}$ of the deformed configuration such that $\mathcal{N}$ contains *P*, there exists a

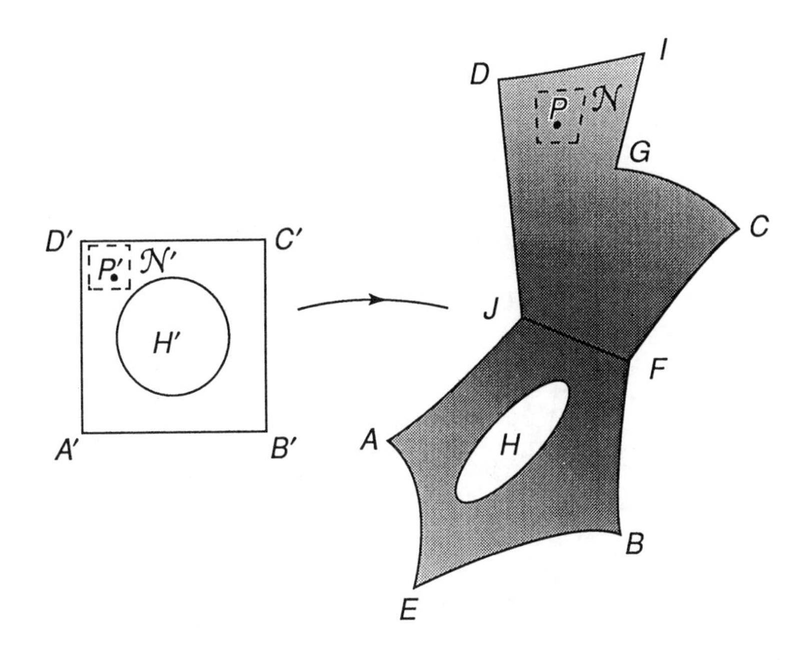

Figure 82 A square with a hole is deformed into a surface with a hole, a bend, and new corners

piece $\mathcal{N}'$ of $A'B'C'D'$, such that $\mathcal{N}'$ contains P' and $\mathcal{N}'$ is deformed into $\mathcal{N}$. This suggests a definition:

Definition (S1): An *elementary surface* is a set $\mathcal{S}$ of points in three-dimensional Euclidean space with the following property: at every point P of $\mathcal{S}$, we can construct a closed ball $\overline{\mathcal{B}}_r(P)$, of radius r (> 0) and center P, such that the intersection $\mathcal{N} = \overline{\mathcal{B}}_r(P) \cap \mathcal{S}$ can be obtained by a deformation of some closed disk $\mathcal{N}'$ belonging to the plane. See Fig. 83.

A few observations should be made regarding Definition (S1). First, it is based on a *local* examination of $\mathcal{S}$ at each of its points. Second, the intersecting ball is taken to be closed, *i.e.,* it contains all points of space located a distance $d \leq r$ from P (in contrast to an open ball

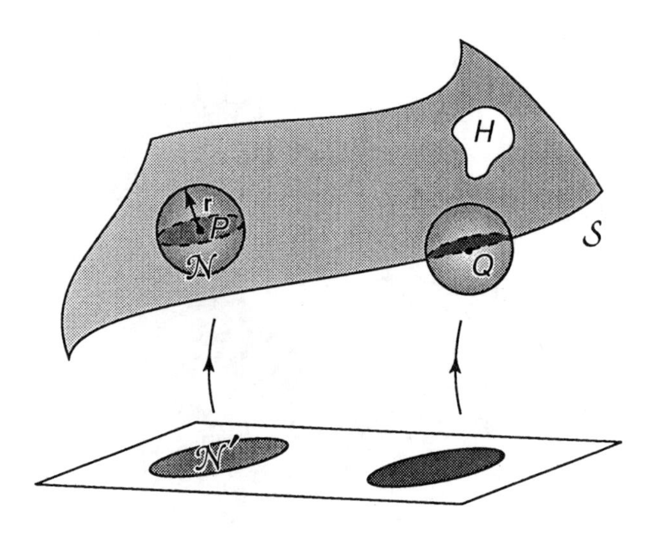

Figure 83 At a point P belonging to $\mathcal{S}$, a closed ball centered at P intersects $\mathcal{S}$ in $\mathcal{N}$. $\mathcal{N}$ is obtained from a closed disk $\mathcal{N}'$ by a deformation.

which excludes the points located at $d = r$). Closed balls can be found at both interior points of $\mathcal{S}$ and points on the boundary (such as Q in Fig. 83).[2] Third, the disk $\mathcal{N}'$ can be replaced by any plane region that can be obtained by a deformation of a disk, *i.e.,* by the closed region surrounded by any simple closed curve. Fourth, the similarity between Definition (S1) and Definition (C1) in Chapter 7 should be noted. Fifth, to check physically whether a given geometrical object satisfies Definition (S1) or not, one can take a small disk of elastic material and try to deform it in such a way that it can be brought into local coincidence with the object. (A small square will also do, since it can always be deformed into a disk.)

In the case of a curve, we referred to a piece cut out by a closed ball as an *arc* (p. 90). We do not have a correspondingly suitable word in the case of a surface, so we will sometimes use the expression "piece of surface" to refer to the intersection made by a closed ball and a surface. A piece of surface can always be obtained through a deformation of a closed disk. A piece of surface is, of course, itself an elementary surface.

It is easy to see that planes, cylinders, spheres, cones, cubes, and polyhedra are all elementary surfaces. So are the surfaces of watermelons,

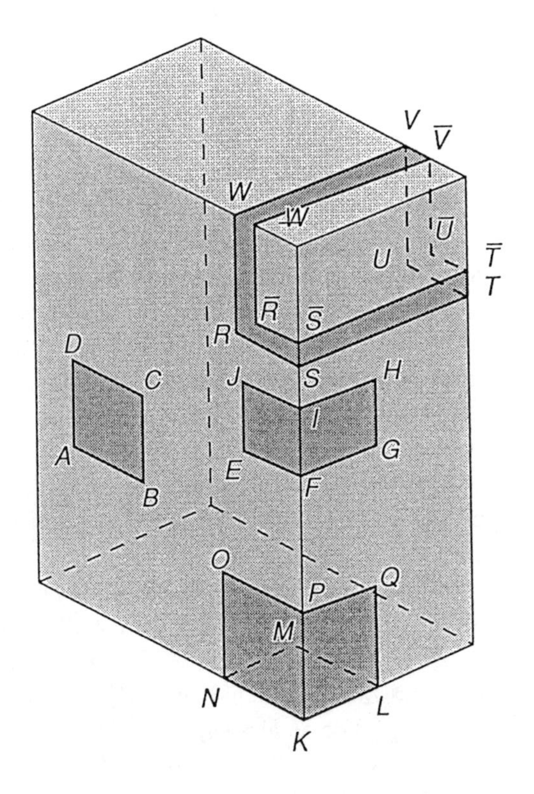

Figure 84 Obtaining a variety of surfaces from a box

eggplants, bowls, cups, liquid-detergent bottles, and the surfaces of a myriad of other things.

Experiment 28 (Making elementary surfaces from a box): Take two identical cardboard boxes, and mark out various pieces on both (some examples are suggested in Fig. 84). Keep one box intact, and cut the marked pieces out of the other. Do the cut pieces constitute elementary surfaces? (Save your material for later experiments.) □

Experiment 29 (More elementary surfaces): Make or find models of cylinders, cones, cubes, various polyhedra,[3] and more general elementary surfaces. □

Consider next the figures in Fig. 85. In Fig. 85a, two squares are joined together by a line IJ. Without the line, each square would qualify as an elementary surface, and the figure formed from the union of the separated squares would also be an elementary surface. But, when the line IJ is included, the resulting figure is not an elementary surface: no piece of IJ can be obtained through a deformation of a disk (more than one point of the disk would have to be mapped into a single point of the line and hence the process cannot consist of a family of homeomorphisms). Similarly, the two triangles touching at I in Fig. 85b do not constitute an elementary surface. Fig. 85c represents a can which has been opened up at the top and bottom. If the top and bottom were completely removed, each of them would be an elementary surface, and so also would the lateral portion of the can. However, as connected in the figure, they do not form an elementary surface due to the impossibility of deforming a disk into the intersection formed by the can and any ball centered at I or J. In Fig. 85d, two planes intersect along IJ. Taken together, these do not form an elementary surface, but if cut along IJ, each of the resulting pieces and their union would be an elementary surface.

Each of the three figures in (b), (c), and (d) of Fig. 85 is the union of two elementary surfaces. Paralleling our Definition (C2) in Chapter 7 for a curve, we can make an analogous definition for surfaces:

Definition (S2): A *surface* is any set that can be regarded as the union of finitely many elementary surfaces.

In accordance with this definition, the figures in (b), (c), and (d) of Fig. 85 are surfaces. The figure in (a) of Fig. 85 is not a surface: it is the union of a line segment and a surface (consisting of two squares).

Experiment 30 (A Möbius strip): Take a long strip of thin cardboard (or of paper) and mark it as in Fig. 15. Hold the end AB fixed. Rotate the end CD a half-turn (180°) either clockwise, or alternatively counter-clockwise, relative to AB, and tape the ends together. This is a model

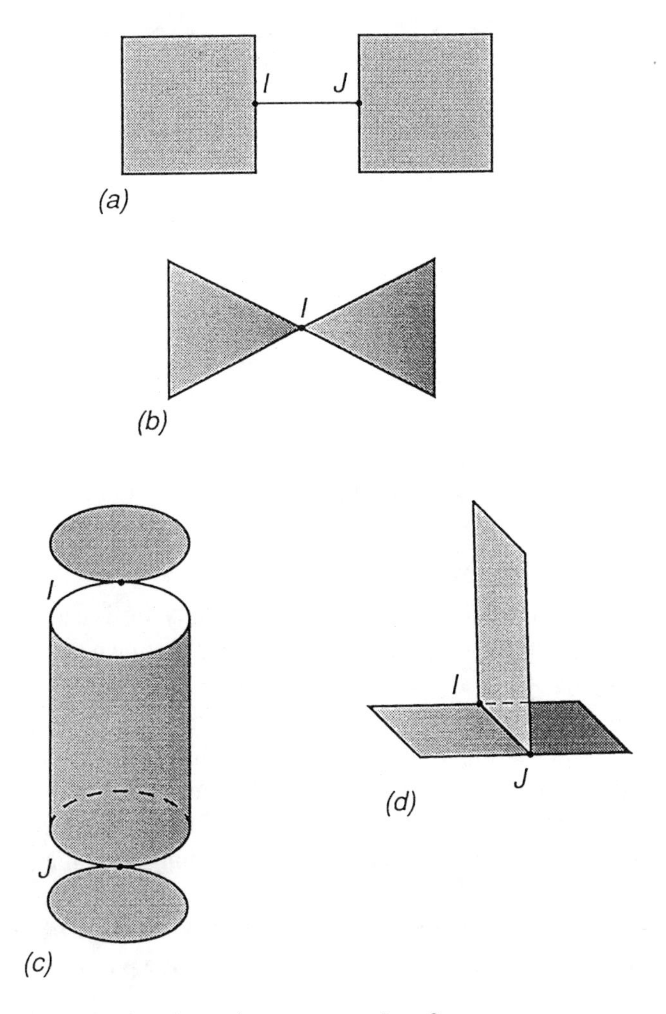

Figure 85 Are these elementary surfaces?

Figure 86 A Möbius strip

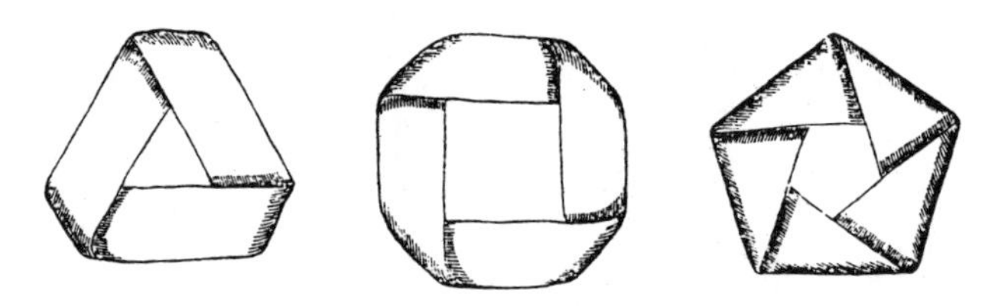

Figure 87 More surfaces by Möbius

of a *Möbius strip* (Fig. 86).[4]

(a) Argue that the Möbius strip is an elementary surface, *i.e.*, that it satisfies Definition (S1) (even though we did not, of course, manufacture our model by means of a topological process).

(b) Draw a line down the center of the strip and keep going until you arrive back at the starting point. What do you find?

(c) Color a small piece of the strip red, turn it over and color the portion of the surface that now faces you black. Turn the strip over again and apply more red. Repeat this process for as long as you can. What happens?

(d) Using a colored marker, start anywhere on an edge of the strip and continue coloring until you return to the starting point. Is a part of an edge left uncolored? Explain.

(e) Cut the strip along the line you drew in **(b)**. Describe what occurs.

(f) Make another Möbius strip and cut it lengthwise along a line that lies one-third of the width from an edge. What happens?

(g) Try cutting Möbius strips in other ways, too.

(h) Do cuts on some of the pieces that you got in **(d), (e)**, and **(f)**.

(i) State several ways in which a Möbius strip differs from a cylindrical strip. □

In Experiment 30, you will have discovered that while a cylindrical strip has two sides (and two edges), the Möbius strip has only one side (and one edge). Locally, the Möbius strip has two sides, but the surface as a whole has only one side.

Experiment 31 (More surfaces by Möbius): The drawings in Fig. 87 are copied from Möbius's collected works.[5] You can make models of them from strips of paper or fabric. Explore the models as in Experiment 30, and you will find some very interesting geometrical phenomena. □

Experiment 32 (Some more surfaces): Cut out a large piece of paper in a shape such as the example shown in Fig. 88. Twist some of the horizontal strips and glue them to one another or to the vertical strip. Glue the two ends of the vertical strip together. Add some curved strips too. Be creative! Describe your findings. □

Experiment 33 (Soap films I): You can generate beautiful surfaces from a mixture of water and liquid dish-washing detergent by the following two methods:[6,7]

(a) Bend a wire to form a closed curve with a handle. Dip it in the mixture and gently remove.

(b) Take two sturdy drinking straws, thread a fairly long string through them, and knot the string to form a closed loop. Pull on the string to obtain a taut loop and dip it in the mixture. Gently remove, and rotate the straws relative to one another to generate a variety of surfaces. (Save the mixture for later experiments.) □

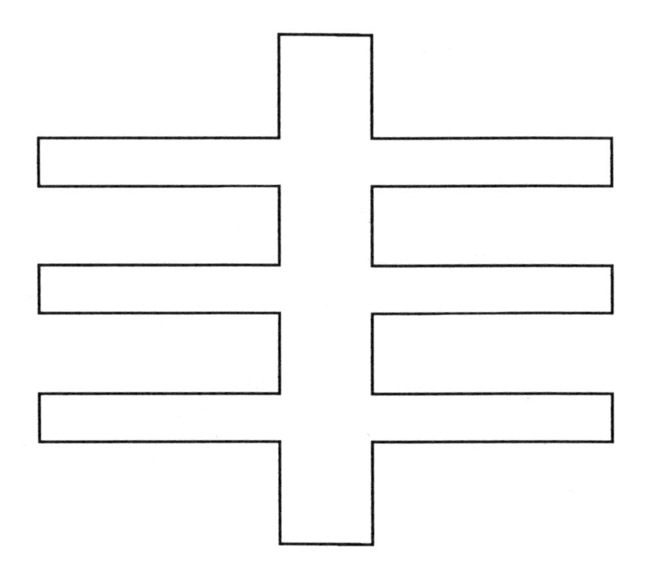

Figure 88 Preparing to make another surface

Connectivity of Surfaces

We would all agree that once the square $ABCD$ in Fig. 84 has been cut out from the cardboard box, it is no longer "connected" to it. Likewise, if the line IJ were removed in Fig. 85a, the two squares would be said to be disconnected from one another. In mathematics, a few distinct concepts of connectivity arise. One useful one is that of arcwise connectedness: a surface $\mathcal{S}$ is *arcwise connected* if and only if each pair of distinct points of $\mathcal{S}$ can be joined by a curve lying entirely in $\mathcal{S}$. The two surfaces $A'B'C'D'$ and $ABCD$ in Fig. 81, as well as both surfaces in Fig. 82, are arcwise connected. If the line IJ is omitted from Fig. 85a, the surface consisting of the union of the two squares is not arcwise connected. Both a cylindrical strip and a Möbius strip are arcwise connected. So is the

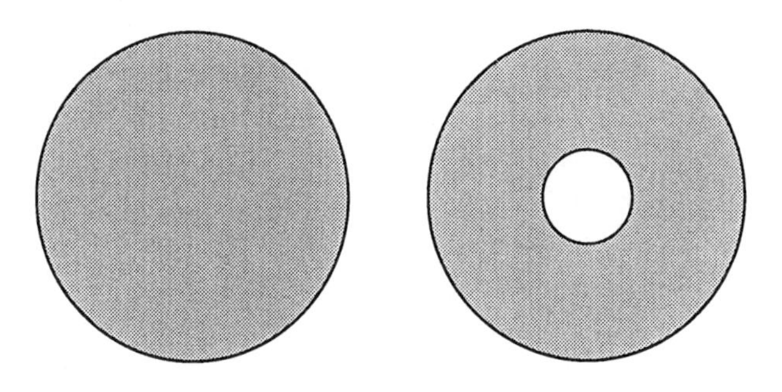

Figure 89 A disk and an annulus

entire plane. The surfaces of spheres, doughnuts, and pretzels are also arcwise connected.

Consider now the two surfaces in Fig. 89. On the left is a disk and on the right is a disk with a hole, *i.e.,* an annulus. Both the disk and the annulus are arcwise connected surfaces. However, the disk is not homeomorphic to the annulus. [By puncturing the disk and deforming it, one could make an annulus; by cutting out a hole, one could also make an annulus. But, these are not topological operations.] The hole of the annulus makes its presence felt also in the following way: Suppose that you take a small rubber band, stretch it and hold it anywhere on the disk, and then let it slowly unstretch. There is nothing to stop it from recovering its original shape, no matter how small it initially was. On the other hand, if you stretch the rubber band to make it surround the hole of the annulus, it cannot recover its original size while still remaining on the annulus. Of course, if you had placed the rubber band on the annulus in such a way that it did not surround the hole, you would not have encountered this difficulty. But, the fact remains, that there *is* a way that the difficulty can arise for an annulus, while there is no way that it can arise for a disk. This leads us to a definition: A surface S is said to be *simply connected* if and only if *every* simple closed curve (*i.e.*, every Jordan curve) in S can be continuously shrunk to a point of S while remaining in S throughout this process. So, the disk is simply connected, whereas the annulus is not. It is clear that the square $A'B'C'D'$ in Fig. 81 is simply connected, as also is the surface $ABCD$. Neither of the surfaces in Fig. 82 is simply connected, since a closed curve that surrounds the hole could not be continuously shrunk to a point of the surface. A spherical surface is simply connected, and the entire surface of a can is. However, the cylindrical surface of a can is not simply connected, because a rubber band girding the can could not be contracted to a point. The surfaces of cubes and tetrahedra are simply connected. The surface of a doughnut (known as a *torus*) is not simply connected, as you can easily see by touching your thumb with your index finger through the hole and by trying to reduce the space formed by your fingers. The surface of a pretzel is not simply connected either.

Although it may sound strange, it is possible for a surface to be simply connected without being arcwise connected. For example, if the annulus in Fig. 89 is replaced by a disk, we would then have two simply connected surfaces in the figure. Moreover, the union of the two would also be simply connected, as you can check from the definition. However, the union of the two disks is certainly not arcwise connected. Thus, even though the terms "arcwise connected" and "simply connected" both contain the word "connected", they actually represent two distinct concepts.

Experiment 34 (Connectivity): (a) Examine the surfaces which you had in Experiments 28-33 to see which are arcwise connected, which are simply connected, which are both, and which are neither.

(b) Look at the mugs, jars, jugs, pots, kettles, tables, chairs, *etc.,* in your house, and determine which have simply connected surfaces and which do not. (One test is to tie a piece of string around the handle of a jug, for example, and try to imagine it shrinking indefinitely.) □

Several times already, we have encountered deformations of surfaces. In Experiment 5, for instance, an elastic strip was stretched out on a table, and in Fig. 81 the deformation of a square into a non-planar, or curved surface, is depicted. Also, looking at Fig. 27, you can imagine how a hemisphere can be deformed into a disk. To do this physically, we would have to choose a highly elastic material and apply appropriate forces to it. But, as explained in Chapter 5, all we need mathematically is a continuous family of homeomorphisms (or topological mappings) which sequentially take the original shape into the final one. In Fig. 90, we indicate how a conical surface may be deformed into a disk. Clearly, then, a conical surface is also deformable into a hemispherical one. In

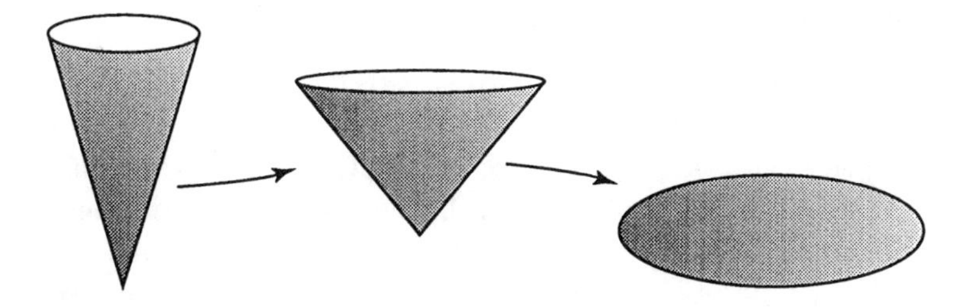

Figure 90 Deformation of a conical surface into a disk

Fig. 91, a cylindrical surface is deformed into an annulus. The surfaces of spheres, ellipsoids, and cubes can all be deformed into one another, and hence are all topologically equivalent. In Fig. 92, a torus is first made to bulge in one place, and is subsequently deformed into a spherical surface with a handle (you should imagine that fluid is able to flow inside the torus, and also from the spherical region into the top of the handle and back through the bottom). Similarly, you can deform the torus into a cubical surface with a handle having flat sides. The surface of a pretzel can be deformed into a sphere with two handles. You can have lots of fun trying to imagine what strange figures can be made by deforming ordinary ones!

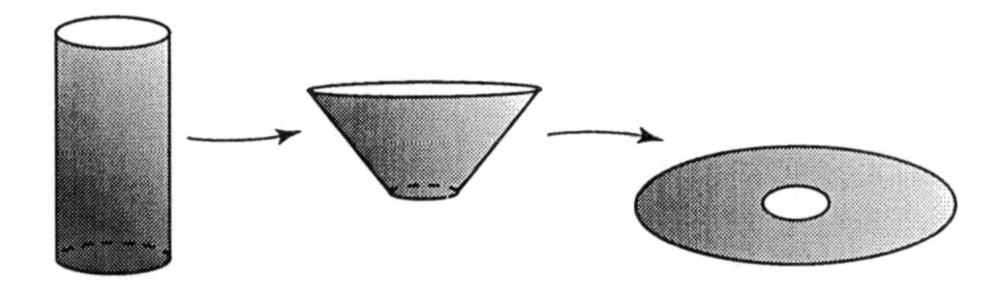

Figure 91 Deformation of a cylinder into an annulus

It is obvious that deformations can severely change the shapes of surfaces and the figures drawn on them. Yet, remarkably, some properties do not change.

Experiment 35 (Deflated figures): (a) Take two balloons, inflate both of them fully, and tie them with strings in such a way that you can later open them. Draw the same selection of figures on both, including some intersecting curves, right angles, a simply connected portion of the surface, and a non-simply connected portion. Deflate one of the balloons partially. Identify which properties have been changed and which have not.

(b) Take the rubber strip that you used in Experiment 5, draw some more figures on it, and stretch it over part of the surface of a bowl or a vase. (An alternative is to use stretchable fabrics with interesting designs). Which properties change and which do not? □

In this experiment, you will have observed that although lengths, angles, shapes, and sizes of figures have changed, some properties have not. Most fundamental of all, the surface of the balloon retains its property of *being a surface* throughout cycles of inflation and deflation. But also, closed curves remain closed curves, and points of intersection of curves continue to be points of intersection.

You will also have observed that simply connected portions of the

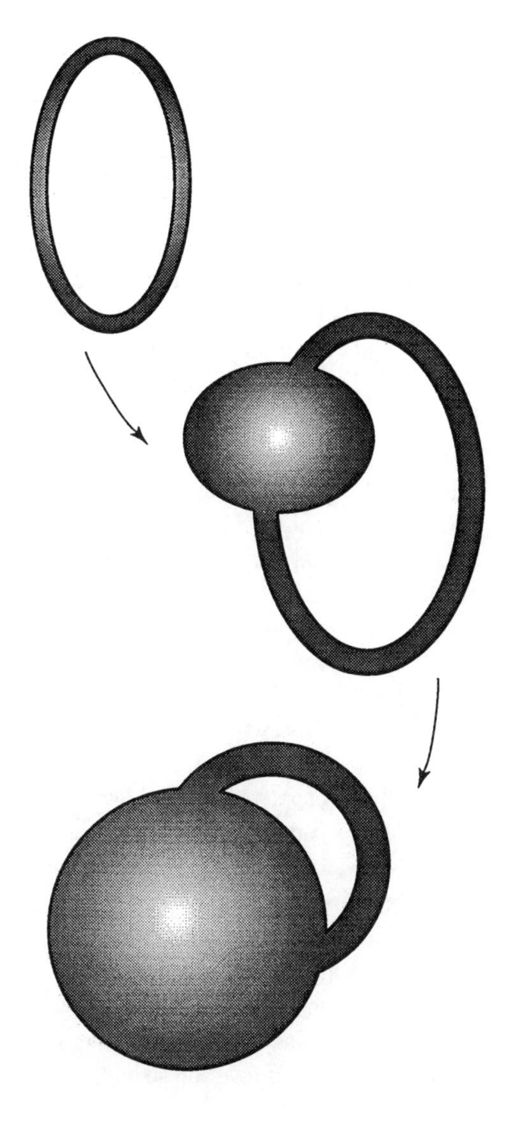

Figure 92 Deformation of a torus into a sphere with a handle

surface remain simply connected, and non-simply connected portions remain non-simply connected. Likewise, arcwise connected portions remain arcwise connected. None of the properties that have been preserved are metrical in character, that is to say, they do not have to do with the measurement of lengths. They belong to the surface in a more deeply rooted way than that.

In Experiment 8, we performed some perfect cut-and-join operations, from which it became clear that two figures can be homeomorphic to one another (or topologically equivalent) without being deformable into one another. Taking another example, we could cut a torus, make a knot in it, and rejoin the ends in such a way that all end points are brought back into coincidence. The initial and final figures are topologically equivalent, but they cannot be deformed into one another. Yet, some properties remain the same: both figures are elementary surfaces, both are arcwise connected, and neither one is simply connected. Any property that is preserved under all topological mappings is called a *topological property*. The topological properties of a figure are its very deepest geometrical properties.

Notes to Chapter 11

[1] Recall the discussions in Chapters 2 and 5. Mathematically, at every stage of a deformation process, the mapping between the undeformed and the deformed configurations is a *homeomorphism* (*i.e.,* a one-to-one correspondence that is continuous and has a continuous inverse). Overlapping brings distinct points into coincidence, while tearing takes points lying on a line into two separate lines. In puncturing, a single point is mapped into a curve. None of these three processes is a homeomorphism. If a square is imagined to be somehow dissected into the infinitely many points that compose it, this is a one-to-one correspondence, but continuity is lost, and so it also is not a homeomorphism.

[2] For some mathematical purposes, especially when doing calculus on surfaces, open balls are employed in the definition of a surface (see B. O'Neill (1966) and M. Spivak (1970)).

[3] Several authors have described the construction of polyhedral models. See, for instance: Wenninger (1966, 1971), Pugh (1976), Hilton and Pedersen (1988), and Lichtenberg (1988).

[4] A.F. Möbius (1790-1868) was a mathematician and astronomer at Leipzig University. He studied theoretical astronomy under Gauss (whose biography is sketched in Chapter 14). He discovered the one-sided strip in 1858. It is described on pp. 484-485 of Vol. 2 of his *Collected Works* (Möbius, 1886), in a paper that was published in 1865. Interestingly, the one-sided strip was discovered independently by J.B Listing, also in 1858 (see M.J. Crowe's entry on Möbius in *The Dictionary of Scientific Biography).* A collection of essays on Möbius's work, set in the context of his times, can be found in Fauvel, Flood, and Wilson (1993).

[5] Fig. 85 is copied from p.520 of Vol. 2 of Möbius's *Collected Works* (Möbius, 1886).

[6] Adding glycerin to the soap mixture will result in longer-lasting bubbles.

[7] Some fascinating (and mathematically important) examples of soap films are discussed in Courant (1940). The classic book on the subject is Boys (1911).

12

Surface Measurements

Having defined what surfaces are and having studied some of their topological properties, we now begin to explore their geometry quantitatively. In the present chapter, we will be concerned primarily with the measurement of distances and angles on surfaces. We will see how such *metrical* properties of surfaces can be expressed in terms of certain fundamental quantities called the metric coefficients. The theory discussed here and in Chapter 13 was invented single-handedly in the early part of the 19th century by the great mathematician Gauss (whose biography is sketched in Chapter 14). It was a major turning point in the history of geometry.

Representing Surfaces Algebraically

In describing curves algebraically (see *e.g.*, Examples 8-10 of Chapter 4), two alternative methods were employed: in one method, an equation is used which specifies implicitly the relationship between the Cartesian coordinates of a point on the curve; in the other method, the Cartesian coordinates are specified in terms of a parameter, such as angle, or time, or arc length. For surfaces also, these two methods prove useful, as the following examples illustrate.

Example 1: The sphere. Let P be a generic point on a sphere of radius R, centered at the origin (Fig. 93). Recall from Example 1 of Chapter 6 that the position vector $\mathbf{r}$ of P can be represented by Equation (6.1) and that the magnitude $||\mathbf{r}||$ of $\mathbf{r}$ is given by Equation (6.2). Since $||\mathbf{r}||$ is the distance of P from O, it is obvious that the equation of the sphere can be written as

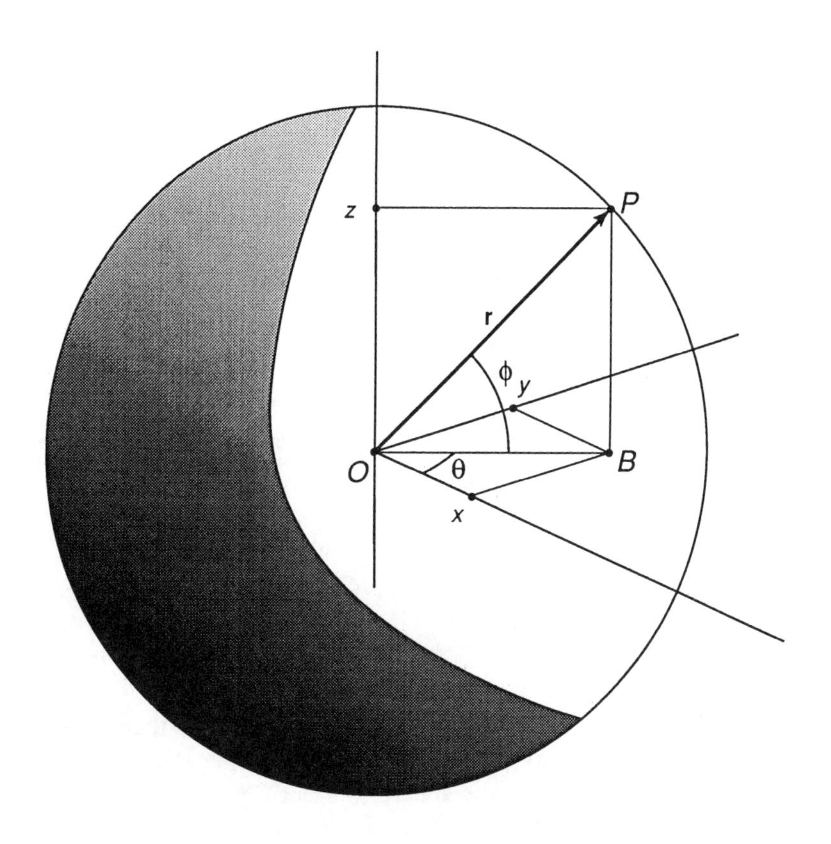

Figure 93 The sphere

$$\sqrt{x^2 + y^2 + z^2} = R \ , \tag{12.1}$$

or equivalently as

$$x^2 + y^2 + z^2 = R^2 \ . \tag{12.2}$$

Equations (12.1) and (12.2) specify the relationship between the coordinates x, y, z implicitly.

Let us introduce angles θ and ϕ as indicated in Fig. 93. Borrowing geographical terminology, these are called the longitude and latitude of P, respectively. Noting that

$$OB = R\cos\phi \ , \ BP = R\sin\phi \ , \qquad (12.3)$$

we find that the rectangular Cartesian coordinates of P can be expressed in terms of θ and ϕ as:

$$\begin{aligned} x &= R\cos\phi\cos\theta \ , \\ y &= R\cos\phi\sin\theta \ , \\ z &= R\sin\phi \ . \end{aligned} \qquad (12.4)$$

Therefore, the position vector of P can be written in the form

$$\mathbf{r} = R\,(\cos\phi\cos\theta\,\mathbf{i} + \cos\phi\sin\theta\,\mathbf{j} + \sin\phi\,\mathbf{k}) = \tilde{\mathbf{r}}(\theta,\phi) \ , \quad (12.5)$$

where the notation $\tilde{\mathbf{r}}$ is an abbreviation for the function that maps (θ, ϕ) into $\mathbf{r}$. The latitude and longitude serve as two parameters for determining the position of P on the sphere. For each θ and ϕ in the intervals $0 \leq \theta < 2\pi$, $-\pi/2 \leq \phi \leq \pi/2$, a unique point on the sphere is specified. Conversely, except for the north and south poles (where $\phi = \pi/2$ and $-\pi/2$, respectively), each point on the sphere is mapped into a unique pair (θ, ϕ) of real numbers; at the poles, the longitude θ can have any value.

Referring again to the point P in Fig. 93, suppose that we fix θ at its value at P, and vary the value of ϕ above and below its value at P, a segment of a great circle passing through the poles will be described (each great circle has O as its center). Likewise, if we fix ϕ and vary θ, a segment of a circle of radius OB will be described on the sphere. If we carry out this procedure for all points in the vicinity of P, two families

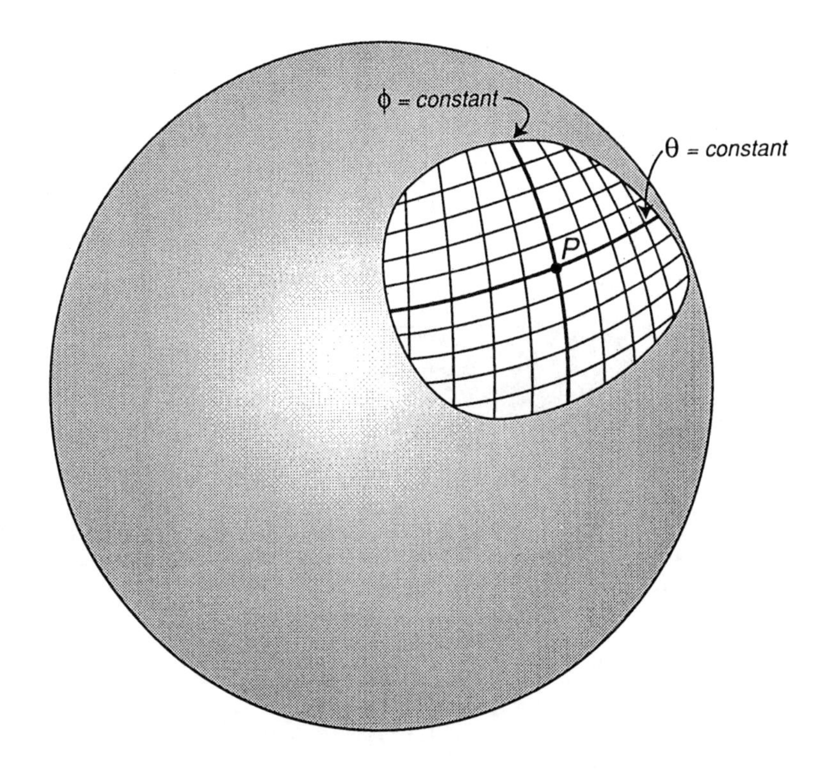

Figure 94 A coordinate net, or chart, on the sphere

of intersecting curves – the meridians and parallels – will be generated. These form a *coordinate net*, or a *chart*, for points close to P (Fig. 94). In the case of the sphere, this chart can be extended to include all points of the sphere except the poles; at each pole, the parallels degenerate to a point and all the meridians meet in this point.

Experiment 36 (A chart): Take a rubber sheet and a globe. Lay the sheet over a portion of the globe and secure it. Trace several segments of lines of latitude and longitude onto the sheet in ink. Untie the sheet and lay it flat. How do the coordinate lines now appear? Try to make the coordinate lines rectangular by pulling on the sheet. □

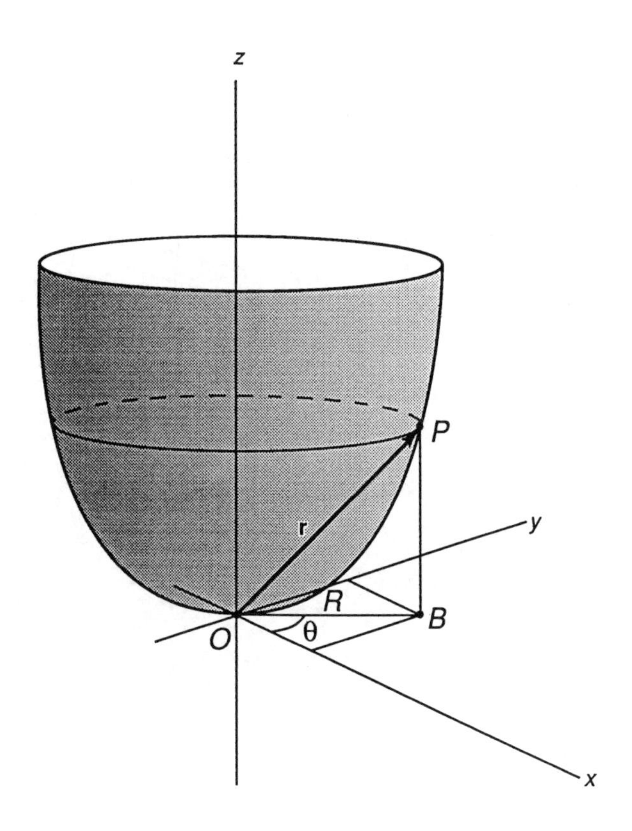

Figure 95 A paraboloid of revolution

Example 2: A paraboloid of revolution. The locus described by the equation

$$z = x^2 + y^2 \tag{12.6}$$

is called a paraboloid of revolution (Fig. 95). Its intersection with the plane $y = 0$ is a parabola $z = x^2$, and all other vertical sections of the surface can be generated by rotating this parabola about the z - axis. The intersection of the paraboloid with a plane $z = R^2$ is a circle of radius R. Equation (12.6) specifies the relationship between the coordinates of a point P on the paraboloid implicitly.

In view of Equation (12.6), we may write the position vector of P as

$$\mathbf{r} = x\,\mathbf{i} + y\,\mathbf{j} + (x^2 + y^2)\mathbf{k} = \hat{\mathbf{r}}(x, y)\,, \tag{12.7}$$

where $\hat{\mathbf{r}}$ is an abbreviation for the fucntion that maps (x, y) into $\mathbf{r}$. Hence, the two coordinates x and y can be used to parametrize the paraboloid. Setting $x =$ constant in Equation (12.7), we find that a parabola is traced out on the surface. Similarly, setting $y =$ constant, another parabola is traced out. These two families of intersecting parabolas form a chart for all points of the paraboloid (Fig. 96).

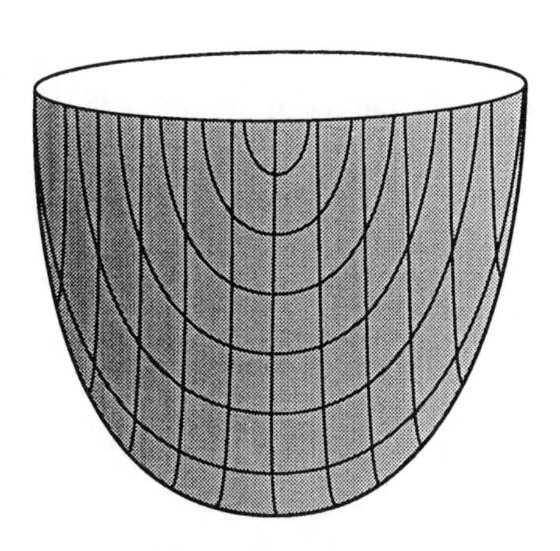

Figure 96 A chart for the paraboloid

Another parametric representation of the paraboloid can be given using the variables R and θ (Fig. 95). Since

$$x = R\cos\theta,\ y = R\sin\theta,\ z = x^2 + y^2\ , \qquad (12.8)$$

we may express the position vector of P as some other function $\tilde{\mathbf{r}}$ of R and θ:

$$\mathbf{r} = R(\cos\theta\,\mathbf{i} + \sin\theta\,\mathbf{j} + R\,\mathbf{k}) = \tilde{\mathbf{r}}(R, \theta)\ . \qquad (12.9)$$

Setting $R =$ constant in Equation (12.9), we find that a circle is described on the paraboloid. Likewise, setting $\theta =$ constant in the Equation (12.9), we obtain the intersection of the paraboloid with a half-plane containing the z-axis and the line OB in Fig. 95, which is a half-parabola. The family of circles and the family of half-parabolas together furnish a chart at each point of the paraboloid, except at the vertex, where the family of circles collapses to a point and θ can have any value.

Example 3: The torus. Suppose that a circle of radius b is rotated about the vertical axis in such a way that its center moves in a circle of radius $a > b$ (Fig. 97): the surface thereby swept out is a torus. Choosing coordinates θ and ϕ $(0 \leq \theta < 2\pi, 0 \leq \phi < 2\pi)$ as indicated in Fig. 97, we may express the position vector of a point P on the torus as

$$\mathbf{r} = (a + b\cos\phi)\,(\cos\theta\,\mathbf{i} + \sin\theta\,\mathbf{j}) + \ b\sin\phi\,\mathbf{k} = \tilde{\mathbf{r}}(\theta,\ \phi)\ . \quad (12.10)$$

Hence,

$$\begin{aligned} x &= (a + b\cos\phi)\cos\theta\ , \\ y &= (a + b\cos\phi)\sin\theta\ , \\ z &= b\sin\phi\ . \end{aligned} \qquad (12.11)$$

Thus, the two coordinates θ, ϕ can be used to parametrize the torus. Clearly, setting $\theta =$ constant yields a vertical circle on the torus, and setting $\phi =$ constant yields a horizontal circle of radius ($a + b\cos\phi$). These two families of intersecting circles furnish a chart for the entire torus. A portion of the chart is shown in Fig. 97.

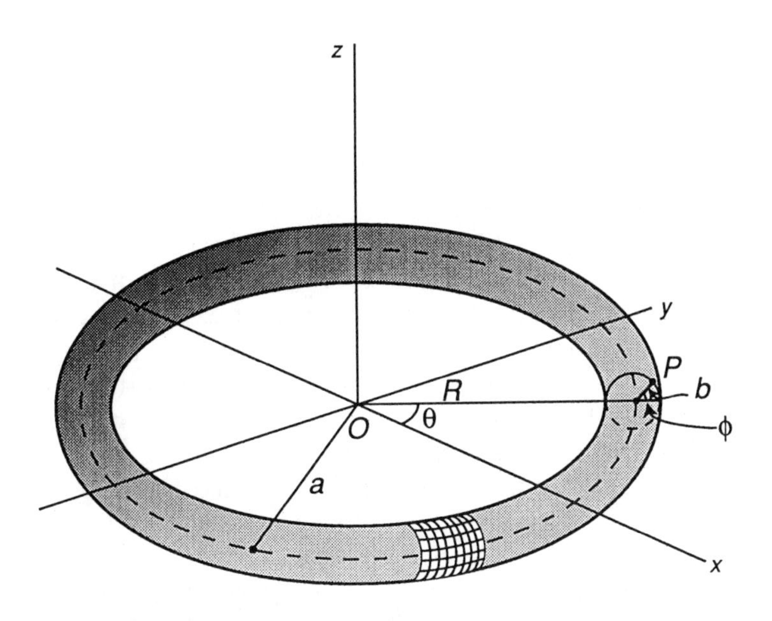

Figure 97 The torus

An implicit equation for the torus can be obtained from Equations (12.11) as follows: Note that since $\cos^2\theta + \sin^2\theta = 1$, we have

$$x^2 + y^2 = (a + b\cos\phi)^2\,, \tag{12.12}$$

and hence

$$\sqrt{x^2 + y^2} - a = b\cos\phi\,. \tag{12.13}$$

Using the relation for z in Equation (12.11) together with Equation (12.13), we obtain

$$(\sqrt{x^2 + y^2} - a)^2 + z^2 = b^2\,. \tag{12.14}$$

In the foregoing examples, one sees how parametric representations can be introduced for some special surfaces. Since this technique is very useful, let us see if it can be done in a more general way.

Recall from Definitions (S1) and (S2) of Chapter 11 that a surface is composed, at most, of finitely many elementary surfaces, and that when an elementary surface $\mathcal{S}$ is intersected with a sufficiently small closed ball (Fig. 83), the resulting portion of the surface can be deformed into a closed disk in the plane (and *vice versa*). On the disk $\mathcal{N}'$, let us put rectangular Cartesian, or indeed any other set of plane coordinates (Fig. 98). Then, upon deformation of the disk back into the portion $\mathcal{N}$ of the surface, the coordinate lines in the plane are deformed into coordinate curves on $\mathcal{S}$. If we denote the coordinates on $\mathcal{N}'$ by u_1 and u_2, then u_1 = constant describes one family of coordinate curves and u_2 = constant describes the other family. Each point in $\mathcal{N}$ has the same numerical

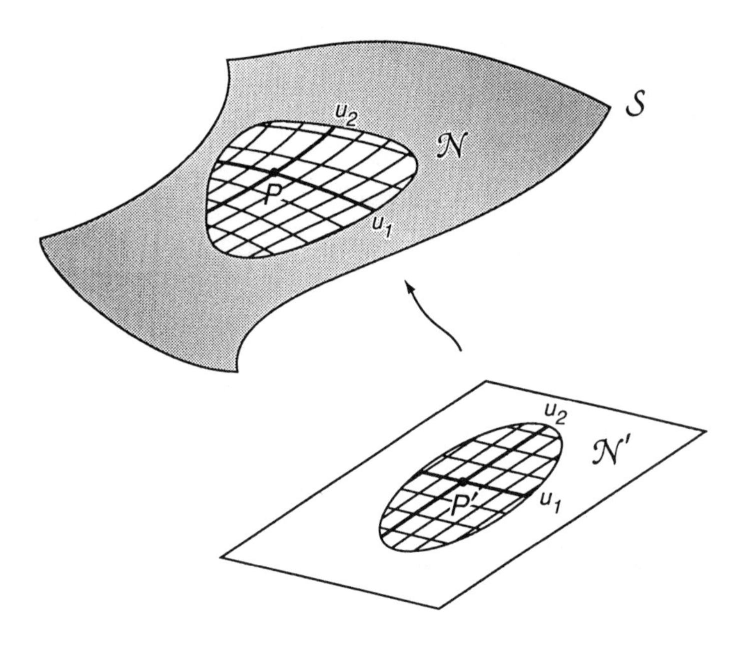

Figure 98 Inducing surface, or Gaussian, coordinates

values of the coordinates (u_1, u_2) as the corresponding point in $\mathcal{N}'$. In the portion $\mathcal{N}$ of $\mathcal{S}$, $u_1 =$ constant describes a family of curves lying on the surface and $u_2 =$ constant describes a second family of curves. The two families of curves intersect one another as indicated in the figure. Through each point of $\mathcal{N}$, there passes one and only one member of each of the two families. Thus, to each point $\mathcal{N}$ there corresponds a unique pair (u_1, u_2) of real numbers, and conversely to each (u_1, u_2), with u_1 and u_2 belonging to specified intervals, there corresponds a unique point of $\mathcal{N}$. In other words, we have established a chart on $\mathcal{N}$. The parameters u_1 and u_2 are called *surface,* or *Gaussian, coordinates*. They provide an ingenious way of identifying points on a surface, and can be regarded as a clever extension of the method introduced by Descartes and Fermat in the 17th century. Naturally, in solving problems, one chooses as simple a chart as possible. But, for theoretical discussions, one leaves the choice of the chart open, and one does not mind if the chart works only for a small portion of surface (for a new one can always be constructed on any adjoining portion).

Once a chart is available, we may write the position vector $\mathbf{r}$ of P as a function $\hat{\mathbf{r}}$ of the surface coordinates (u_1, u_2):

$$\mathbf{r} = \hat{\mathbf{r}}(u_1, u_2)\,. \tag{12.15}$$

Equivalently, we may express the Cartesian coordinates x, y, z of P as functions $\hat{x}, \hat{y}, \hat{z}$, respectively, of the surface coordinates:

$$\begin{aligned} x &= \hat{x}\,(u_1, u_2)\,, \\ y &= \hat{y}\,(u_1, u_2)\,, \\ z &= \hat{z}\,(u_1, u_2)\,. \end{aligned} \tag{12.16}$$

A re-examination of Examples 1, 2, 3 reveals how the choices of surface coordinates which we made for the sphere, paraboloid of revolution, and torus, fit into this general scheme.

Experiment 37 (Surface coordinates): (a) Take some squared paper (or draw a rectangular Cartesian grid on a piece of paper). Lay the sheet snugly against a cylindrical container to induce surface coordinates on the cylinder. Turn the paper in different directions to obtain a few different charts.

(b) Draw a rectangular Cartesian grid on a piece of rubber sheet. Stretch thc sheet so that it lays snugly against a portion of the surface of a bowl or vase. Study the variety of surface-coordinate systems that you can make.

(c) Draw a chart on a portion of a surface of some object.

(d) Use a sharp knife to inscribe a chart on the surface of a large potato, or apple, or watermelon.

(e) Lay a network of thread or string across the surface of the object to form a chart. □

Tangent Plane

In Chapter 9, we discussed in detail the concept of a tangent to a

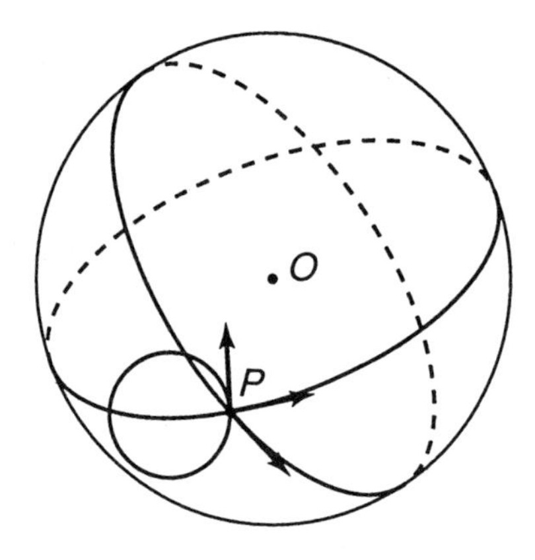

Figure 99 Two great circles and one small circle lying on a sphere and passing through P

curve. If we now take a curve lying on a surface, we can directly apply the ideas of that chapter to it. Thus, for example, each circle drawn on a sphere has a tangent field just like that sketched in Fig. 53. What if we take two great circles passing through a point P on a sphere and another small circle that lies on the sphere and also passes through P (Fig. 99)? Each of these circles has a tangent vector P. If we construct the plane through A that contains the tangents to the two great circles, can we say anything definite about where the tangent vector to the smaller circle lies in relation to this plane?

As another example, consider the point A on the edge BC of the two intersecting rectangles in Fig. 100. As we pass through A going along

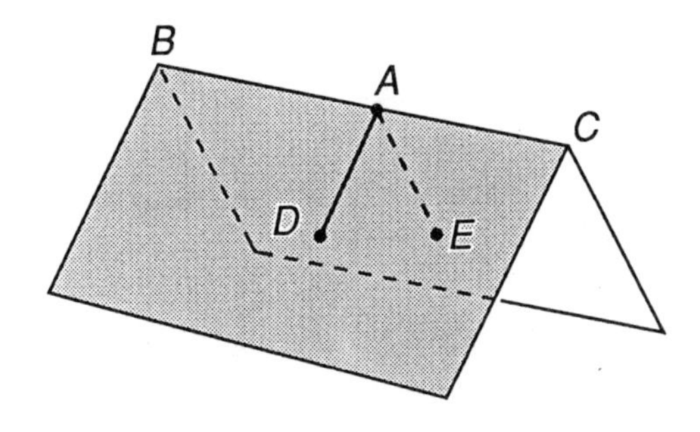

Figure 100 At A, there is a tangent to BAC but not to DAE.

BC, a tangent exists at A. On the other hand, if we go through A along DE, there is a corner at A. A more severe situation occurs at the points B and C in Fig. 100, and a still more severe one at the vertex of a cone.

New light can be shed on the foregoing phenomena by the introduction of the concept of a tangent plane. To this end, consider any point P on a given surface $\mathcal{S}$, and let $TUVW$ be any plane passing through P (Fig. 101). Let PQ be the secant from P to any other point Q on $\mathcal{S}$. Drop a perpendicular QM from Q to the plane $TUVW$, and join PM. The plane $TUVW$ is said to be a *tangent plane* to S at P if and only if it possesses the following property: for every choice of Q, the angle QPM approaches zero as Q approaches P.

It is easy to see that there can be at most one tangent plane at a point on a surface. But, there might not be any. When there is one, all secants

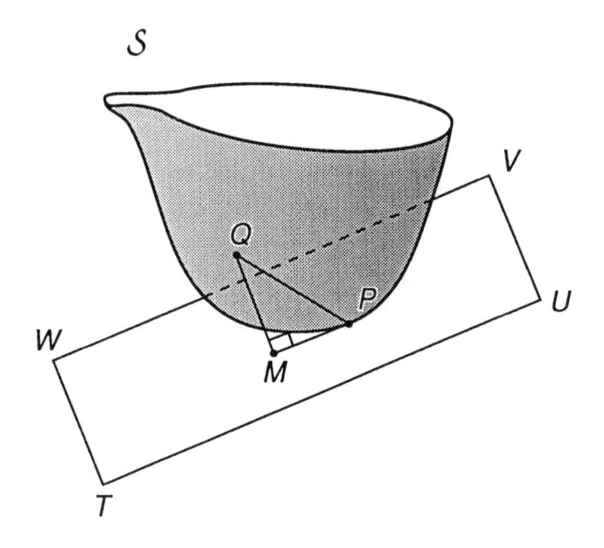

Figure 101 The tangent plane to the surface $\mathcal{S}$ at a point P

PQ fall into the tangent plane at P, as Q approaches P. Further, the tangent vectors to curves lying on $\mathcal{S}$ and passing through P all lie in the tangent plane at P. Thus, in Fig. 99, the three tangent vectors drawn at P must all lie in the same plane. In Fig. 100, there is no tangent plane to the surface at any point along the edge BC. At the edges and corners of a cube, and at the vertex of a cone, there is no tangent plane.

A surface (or a subset of it that is also a surface) is said to be *smooth* if it possesses a tangent plane at each of its points, and if additionally, the tangent plane changes continuously as the piece is traversed.

A line drawn through P and perpendicular to the tangent plane $TUVW$ (Fig. 101) is called a *normal* to the surface at P. Later (in Chapter 15), we will associate a unit vector with this line.

Experiment 38 (Tangent plane): Take some objects such as cylindrical containers, bowls, and vases, and a small square of cardboard.

Use the square to visualize the tangent plane at points on these surfaces where a tangent plane exists. Find points for which the following statements hold: **(a)** the surface meets the tangent plane only at the point of tangency and lies only on one side of the tangent plane.

(b) a straight line in the surface lies in the tangent plane.

(c) the surface lies in the tangent plane.

(d) the surfaces lies on both sides of the tangent plane.

(e) a curve in the surface lies in the tangent plane. □

Surface-Distance

Consider a curve *PRQ* of length l, lying in an arcwise connected surface such as $\mathcal{S}$ in Fig. 102. Let the length of the chord PQ be l_0. In general, this chord does not, of course, lie on the surface. In space, l_0 measures the (Euclidean) *distance* between P and Q. We did not yet define a concept of *distance on a surface.* However, if we consider P to be fixed and let l approach zero, the quotient l_0/l approaches unity. Thus, we can define a measure of distance at surface points "infinitesimally close" together, and for such points the measure is the Euclidean one. Whenever two points on a surface can be joined by a curve, they can be joined by infinitely many curves. In Fig. 102, several different curves that join Q to P are shown. Clearly, not all of these will have the same length. So, how do we define surface-distance between points that are not infinitesimally close to one another? Here is an idea: Let P and Q be any two points on the surface $\mathcal{S}$, and consider the set of all curves on $\mathcal{S}$ that join P to Q. We define the *surface-distance* between P and Q to be the least, or more precisely, the greatest lower bound (infimum), of the lengths of these curves. As an example, consider P and Q to be at opposite ends of a diameter on a sphere of radius 1: the Euclidean distance between them is 2, while the surface-distance is π. We note that since the surface-distance between two points can be changed by a deformation of the surface, surface-distance is not a topological property.

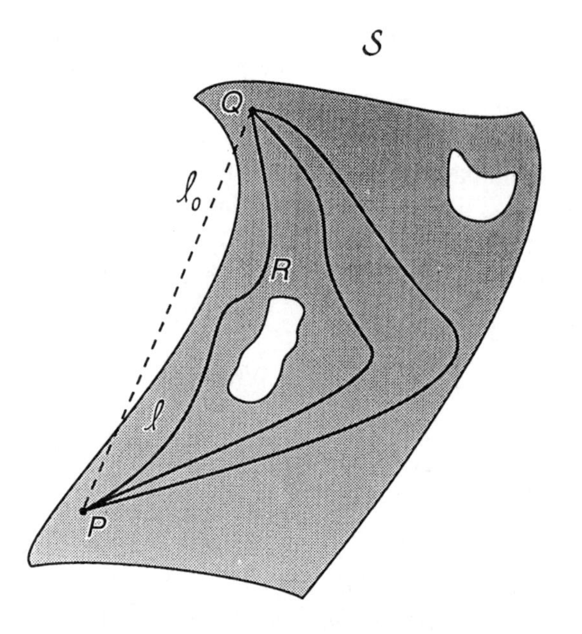

Figure 102 What is the distance between P and Q, measured on the surface $ABCD$?

Experiment 39 (Surface-distance): (**a**) For the surfaces in Experiments 28, 29, and 30, measure the surface-distance between various points as accurately as you can and compare it with the Euclidean distance.

(**b**) Take some polyhedra, cylinders, cones, and other surfaces that you had in Experiment 29. Measure the surface-distance between selected points on them.

(**c**) Find some interesting surfaces on household objects (*e.g.*, a mixing bowl, a vase, a detergent bottle with a handle) and measure some surface-distances on them. □

We now proceed to obtain a very important formula for the distance between two surface points that are infinitesimally close to one another.

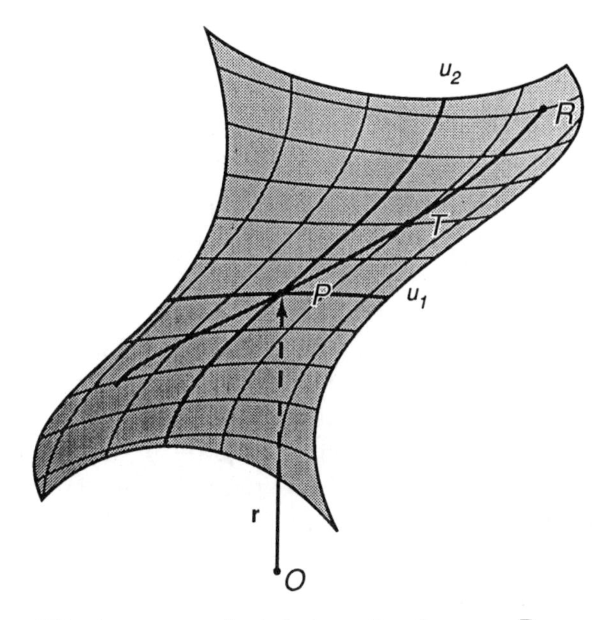

Figure 103 An arc on a charted piece of surface near P

Let us take a piece of surface in the vicinity of the point P in Fig. 102, and let us inscribe a chart on it (Fig. 103). Further, let us parametrize the arc containing P and R by its arc length. Let s be the value of the arc length at P, and let (u_1, u_2) be the surface coordinates of P. Since each value of arc length identifies a unique point on the surface in Fig. 103, and since each such point corresponds to a unique pair of surface coordinates, it follows that the position vector of points along the arc can be expressed as functions of arc length. Thus, remembering Equation (12.15), we may write the position vector of P as

$$\mathbf{r} = \hat{\mathbf{r}}\,(u_1(s), u_2(s)) \;=\; \tilde{\mathbf{r}}\,(s)\;, \tag{12.17}$$

where the notation $\tilde{\mathbf{r}}$ stands for a function that maps arc length into position vectors. Likewise, the rectangular Cartesian coordinates of P

can be expressed in the form

$$
\begin{aligned}
x &= \hat{x}\,(u_1(s),\, u_2(s)) &= \tilde{x}(s)\,, \\
y &= \hat{y}\,(u_1(s),\, u_2(s)) &= \tilde{y}(s)\,, \\
z &= \hat{z}\,(u_1(s),\, u_2(s)) &= \tilde{z}(s)\,.
\end{aligned} \tag{12.18}
$$

Suppose next that on the arc we take a point T close to P. The arc length coordinate of T will be $s + \Delta s$, where Δs is a small increment; similarly, the surface coordinates of T will be $(u_1 + \Delta u_1, u_2 + \Delta u_2)$, and its rectangular Cartesian coordinates will be $(x + \Delta x,\ y + \Delta y,\ z + \Delta z)$. We have seen previously that the increment of arc length is given by the approximate Equation (8.3). Moreover, it was remarked that Equation (8.7) is regarded as an expression for the length of an infinitesimal arc. Keeping these two equations in mind, let us try to develop a formula for dx for the case of an arc lying on a surface. For this case, we have, by virtue of the first of Equations (12.18),

$$
\begin{aligned}
\Delta x &= \tilde{x}(s + \Delta s) - \tilde{x}(s) \\
&= \hat{x}(u_1(s + \Delta s), u_2(s + \Delta s)) - \hat{x}(u_1(s), u_2(s))\,.
\end{aligned} \tag{12.19}
$$

Adding and subtracting the term $\hat{x}(u_1(s), u_2(s + \Delta s))$, we may write Equation (12.19) as

$$
\begin{aligned}
\Delta x &= \hat{x}(u_1(s + \Delta s), u_2(s + \Delta s)) - \hat{x}(u_1(s), u_2(s + \Delta s)) \\
&\quad + \hat{x}(u_1(s), u_2(s + \Delta s)) - \hat{x}(u_1(s), u_2(s)) \\
&= \hat{x}(u_1 + \Delta u_1,\ u_2 + \Delta u_2)) - \hat{x}(u_1, u_2 + \Delta u_2)) \\
&\quad + \hat{x}(u_1, u_2 + \Delta u_2) - \hat{x}(u_1, u_2)\ .
\end{aligned} \tag{12.20}
$$

Next, dividing by Δs and writing $\Delta s = (\Delta s/\Delta u_1)\Delta u_1 = (\Delta s/\,\Delta u_2)\,\Delta u_2$, (for $\Delta u_1, \Delta u_2$ not zero), we get

$$\frac{\Delta x}{\Delta s} = \frac{\hat{x}(u_1 + \Delta u_1, u_2 + \Delta u_2) - \hat{x}(u_1, u_2 + \Delta u_2)}{\Delta u_1} \frac{\Delta u_1}{\Delta s} + \frac{\hat{x}(u_1, u_2 + \Delta u_2) - \hat{x}(u_1, u_2)}{\Delta u_2} \frac{\Delta u_2}{\Delta s} . \tag{12.21}$$

If we now let $\Delta s \to 0$, recall from calculus that the various quotients in Equation (12.21) become derivatives and partial derivatives, provided that all the relevant functions are sufficiently well behaved, which we assume to be the case. Thus, Equation (12.21) becomes

$$\frac{dx}{ds} = \frac{\partial x}{\partial u_1} \frac{du_1}{ds} + \frac{\partial x}{\partial u_2} \frac{du_2}{ds} . \tag{12.22}$$

Here, $\partial x / \partial u_1$ and $\partial x / \partial u_2$ represent partial derivatives of the function $\hat{x}$ in Equations (12.16) with respect to the variables u_1 and u_2. Formally, we may multiply both sides of Equations (12.22) by ds to obtain

$$dx = \frac{\partial x}{\partial u_1} du_1 + \frac{\partial x}{\partial u_2} du_2 . \tag{12.23}$$

Thus, the infinitesimal increment dx is expressible in terms of the infinitesimal increments du_1, du_2 in the surface coordinates and the partial derivatives $\partial x / \partial u_1$, $\partial x / \partial u_2$. Similarly,

$$\begin{aligned} dy &= \frac{\partial y}{\partial u_1} du_1 + \frac{\partial y}{\partial u_2} du_2 \\ dz &= \frac{\partial z}{\partial u_1} du_1 + \frac{\partial z}{\partial u_2} du_2 . \end{aligned} \tag{12.24}$$

From Equations (8.7), (12.23), and (12.24), it follows at once that

$$
\begin{aligned}
ds^2 &= dx^2 + dy^2 + dz^2 \\
&= \left(\frac{\partial x}{\partial u_1} du_1 + \frac{\partial x}{\partial u_2} du_2\right)^2 \\
&+ \left(\frac{\partial y}{\partial u_1} du_1 + \frac{\partial y}{\partial u_2} du_2\right)^2 \\
&+ \left(\frac{\partial z}{\partial u_1} du_1 + \frac{\partial z}{\partial u_2} du_2\right)^2 .
\end{aligned} \tag{12.25}
$$

Multiplying out the terms in parentheses and introducing the abbreviations

$$
\begin{aligned}
a_{11} &= \left(\frac{\partial x}{\partial u_1}\right)^2 + \left(\frac{\partial y}{\partial u_1}\right)^2 + \left(\frac{\partial z}{\partial u_1}\right)^2 , \\
a_{12} &= \frac{\partial x}{\partial u_1}\frac{\partial x}{\partial u_2} + \frac{\partial y}{\partial u_1}\frac{\partial y}{\partial u_2} + \frac{\partial z}{\partial u_1}\frac{\partial z}{\partial u_2} = a_{21} , \\
a_{22} &= \left(\frac{\partial x}{\partial u_2}\right)^2 + \left(\frac{\partial y}{\partial u_2}\right)^2 + \left(\frac{\partial z}{\partial u_2}\right)^2 ,
\end{aligned} \tag{12.26}
$$

we may write Equation (12.25) as

$$
ds^2 = a_{11}\, du_1^2 + 2a_{12}\, du_1 du_2 + a_{22}\, du_2^2 . \tag{12.27}
$$

This is the most important formula in surface geometry and was presented by Gauss in 1827.[1] It expresses the distance between two infinitesimally close points on the surface in terms of surface coordinates. In infinitesimal neighborhoods, the geometry of a surface is Euclidean, even though the geometry of small finite neighborhoods is, in general, not so. Thus, a surface can be regarded as an infinite collection of Euclidean spaces that are smoothly joined together. Another way of thinking about this, is to regard the surface as the envelope of its tangent planes (just as circle can be imagined as being formed as an envelope of its tangents). The expression appearing on the right-hand side of Equation (12.27) is called the *first fundamental form.*[2] The functions of $a_{11},\ a_{12},\ a_{22}$ are called the *metric coefficients*; in general, these vary from point to point as one moves across the surface.

Let us suppose for the moment that both the shape of a surface and its location in space are known, as was the case in Examples 1,2,3. Further, let a chart be chosen on a piece of this surface and assume that the functions $\hat{x}(u_1, u_2),\ \hat{y}(u_1, u_2),\ \hat{z}(u_1, u_2)$ are sufficiently nice as to admit the mathematical operations that we have been discussing. It is then evident from Equation (12.26) that the metric coefficients a_{11}, a_{12}, a_{22} can be calculated from the coordinate functions $\hat{x}, \hat{y}, \hat{z}$.

Example 4: Calculating the metric coefficients. (a) Consider a horizontal plane lying in three-dimensional Euclidean space, and charted by rectangular Cartesian coordinates x and y. The equation of this plane is $z =$ constant, and we can choose on it the chart $u_1 = x, u_2 = y$. Then,

$$\begin{aligned} \frac{\partial x}{\partial u_1} &= 1, \quad \frac{\partial x}{\partial u_2} = 1\,, \\ \frac{\partial y}{\partial u_1} &= 0, \quad \frac{\partial y}{\partial u_2} = 1\,, \\ \frac{\partial z}{\partial u_1} &= 0, \quad \frac{\partial z}{\partial u_2} = 0\,. \end{aligned} \tag{12.28}$$

Substituting these values into Equations (12.26), we see immediately that

$$a_{11} = 1\,,\ a_{12} = 0\,,\ a_{22} = 1\,. \tag{12.29}$$

Thus, the metric coefficients are constants for the plane. Moreover, Equation (12.27) reduces to its simplest form:

$$ds^2 = dx^2 + dy^2 \,. \tag{12.30}$$

If x and y are specified as functions of a parameter t, thereby describing a curve in the plane, then the length of the curve can be found by integrating the expression for ds that one obtains from Equation (12.30). For a straight line, x and y can be specified parametrically as in Equations (4.3) and (4.4). Along the straight line, the differentials dx and dy are therefore given by

$$dx = dt \quad , \quad dy = m\ dt \ , \tag{12.31}$$

m being the slope of the line (See Fig. 11). Inserting these relations into Equation (12.30), we find that

$$ds = \sqrt{1 + m^2}\ dt \tag{12.32}$$

Let P_1 and P_2 be any two points on the straight line, and let their coordinates be (x_1, y_1) and (x_2, y_2), respectively. Also, let $t = t_1, s = s_1$ at P_1, and $t = t_2, s = s_2$ at P_2. The distance between P_1 and P_2 is then given by the integral

$$\begin{aligned} s_2 - s_1 &= \int_{P_1 P_2} ds = \int_{t_1}^{t_2} \sqrt{1 + m^2}\ dt \\ &= \sqrt{1 + m^2}\ (t_2 - t_2) \\ &= \sqrt{(x_2 - x_1)^2 + (y_2 - y_1)^2} \,. \end{aligned} \tag{12.33}$$

In other words, for the Euclidean plane, the infinitesimal Pythagorean formula (12.30) integrates to the usual Pythagorean results for distance between a pair of points that are not necessarily infinitesimally close together.

(b) For the sphere (Example 1), θ and ϕ are surface coordinates (except at the poles). Taking partial derivatives of the functions in Equations (12.4), we obtain

$$\begin{aligned}
\frac{\partial x}{\partial \theta} &= -R\cos\phi\sin\theta \quad , & \frac{\partial x}{\partial \phi} &= -R\sin\phi\cos\theta \, , \\
\frac{\partial y}{\partial \theta} &= R\cos\phi\cos\theta \quad , & \frac{\partial y}{\partial \phi} &= -R\sin\phi\sin\theta \, , \\
\frac{\partial z}{\partial \theta} &= 0 \quad , & \frac{\partial z}{\partial \phi} &= R\cos\phi \, .
\end{aligned} \tag{12.34}$$

Substituting these expressions into Equations (12.26), and making use of the trigonometric identity $\sin^2\theta + \cos^2\theta = 1$ (and the same one for ϕ), we find that

$$a_{11} = R^2\cos^2\phi \, , \;\; a_{12} = 0 \, , \;\; a_{22} = R^2 \, . \tag{12.35}$$

Hence, Equation (12.27) becomes

$$ds^2 = R^2\cos^2\phi \, d\theta^2 + R^2 d\phi^2 \, . \tag{12.36}$$

This is the expression for the square of the length of an infinitesimal line element on the sphere. Note that the coefficient a_{11} varies with latitude. The arc length of a curve on the sphere can be gotten by integrating the expression for ds along the curve.

(c) For the torus (Example 3), the Cartesian coordinates are given in terms of surface coordinates θ, ϕ by Equations (12.11). Taking partial derivatives, we get

$$
\begin{aligned}
\frac{\partial x}{\partial \theta} &= -(a + b\cos\phi)\sin\theta\,, & \frac{\partial x}{\partial \phi} &= -b\sin\phi\cos\theta\,,\\
\frac{\partial y}{\partial \theta} &= (a + b\cos\phi)\cos\theta\,, & \frac{\partial y}{\partial \phi} &= -b\sin\phi\sin\theta\,,\\
\frac{\partial z}{\partial \theta} &= 0\,, & \frac{\partial z}{\partial \phi} &= b\cos\phi\,.
\end{aligned}
\tag{12.37}
$$

Hence,

$$
a_{11} = (a + b\cos\phi)^2\,,\quad a_{12} = 0\,,\quad a_{22} = b^2\,, \tag{12.38}
$$

and

$$
ds^2 = (a + b\cos\phi)^2 d\theta^2 + b^2 d\phi^2\,. \tag{12.39}
$$

Experiment 40 (Line element of a globe): (a) Use thread and a ruler to find the interval on the equator that is intercepted by two meridians 30° apart on a globe. Similarly, for latitude 45°, measure the intercept between the two meridians along the line of latitude. Calculate the ratio of the two intercepts. Now, apply Equation (12.36), with $d\phi = 0$, to find the theoretical ratio between the two intercepts. Do your answers agree with one another?

(b) Use Equation (12.36) to estimate the distance between London and Paris. (For the radius of the earth, use the value 6371 km.) □

Continuing our discussion of the first fundamental form, we note that for a sphere, the square of the length of a line element is still given by Equation (12.36) even if the sphere is translated to a different location in space, and even if it is rotated rigidly. This raises the question of whether we can determine the metric coefficients of a surface without knowing anything about the absolute location of the surface in space. In other words, can we determine a_{11}, a_{12}, a_{22} from surface measurements alone? To explore this question, let us try to understand the metric coefficients a little more deeply.

Consider once again the small arc PT in Fig. 103. Recall that the

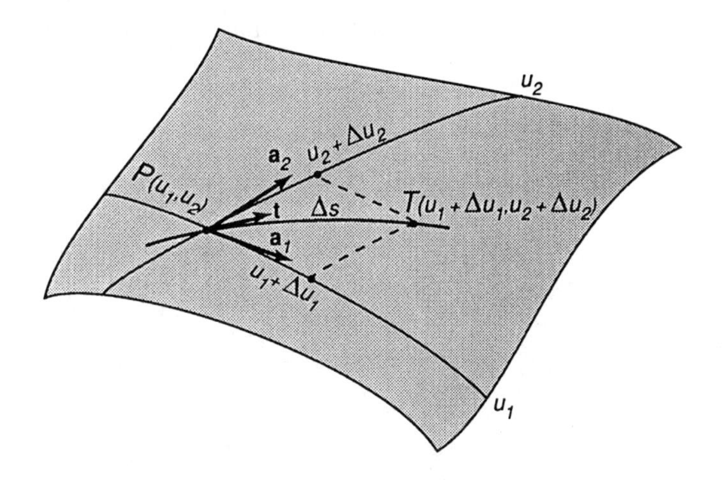

Figure 104 Tangent vectors to the surface-coordinate curves

surface coordinates of P are (u_1, u_2), while those of T are $(u_1+\Delta u_1, u_2+\Delta u_2)$. The four coordinate lines $u_1 = \text{constant}$, $u_2 = \text{constant}$, $u_1+\Delta u_1 = \text{constant}$, and $u_2+\Delta u_2 = \text{constant}$ form a curvilinear quadrilateral figure on the surface (Fig. 104). The position vector of P may be written as

$$\mathbf{r} = \hat{\mathbf{r}}(u_1, u_2) = \hat{x}\,(u_1, u_2)\,\mathbf{i} + \hat{y}\,(u_1, u_2)\,\mathbf{j} + \hat{z}\,(u_1, u_2)\,\mathbf{k}\,. \quad (12.40)$$

Let us denote the vector joining P to T by $\Delta\mathbf{r}$, and let the corresponding infinitesimal increment be $d\mathbf{r}$. We then have

$$\begin{aligned} \Delta\mathbf{r} &= \Delta x\,\mathbf{i} + \Delta y\,\mathbf{j} + \Delta z\,\mathbf{k}\,, \\ d\mathbf{r} &= dx\,\mathbf{i} + dy\,\mathbf{j} + dz\,\mathbf{k}\,, \end{aligned} \quad (12.41)$$

and the unit tangent vector to the arc PT at T is

$$\mathbf{t} = \frac{d\mathbf{r}}{ds} = \lim_{\Delta s \to 0} \frac{\Delta \mathbf{r}}{\Delta s} . \tag{12.42}$$

Now, the quantities dx, dy, dz are related to the infinitesimal increments du_1, du_2 in the surface coordinates by Equations (12.23) and (12.24). We therefore have

$$\begin{aligned} d\mathbf{r} = &(\frac{\partial x}{\partial u_1} du_1 + \frac{\partial x}{\partial u_2} du_2)\mathbf{i} + (\frac{\partial y}{\partial u_1} du_1 + \frac{\partial y}{\partial u_2} du_2)\mathbf{j} \\ &+ (\frac{\partial z}{\partial u_1} du_1 + \frac{\partial z}{\partial u_2} du_2)\mathbf{k} . \end{aligned} \tag{12.43}$$

Introducing two vectors

$$\begin{aligned} \mathbf{a}_1 &= \frac{\partial x}{\partial u_1}\mathbf{i} + \frac{\partial y}{\partial u_1}\mathbf{j} + \frac{\partial z}{\partial u_1}\mathbf{k} = \frac{\partial \hat{\mathbf{r}}}{\partial u_1} , \\ \mathbf{a}_2 &= \frac{\partial x}{\partial u_2}\mathbf{i} + \frac{\partial y}{\partial u_2}\mathbf{j} + \frac{\partial z}{\partial u_2}\mathbf{k} = \frac{\partial \hat{\mathbf{r}}}{\partial u_2} , \end{aligned} \tag{12.44}$$

we see that $d\mathbf{r}$ can be written as

$$d\mathbf{r} = du_1\, \mathbf{a}_1 + du_2\, \mathbf{a}_2 . \tag{12.45}$$

Likewise, the unit tangent vector $\mathbf{t}$ can be expressed as

$$\mathbf{t} = \frac{du_1}{ds}\mathbf{a}_1 + \frac{du_2}{ds}\mathbf{a}_2 . \tag{12.46}$$

What are $\mathbf{a}_1$ and $\mathbf{a}_2$? Well, it is clear from Equations (12.44) that $\mathbf{a}_1$ is gotten by holding u_2 fixed and watching the position vector $\mathbf{r}$ change, and that $\mathbf{a}_2$ is gotten by holding u_1 fixed and watching $\mathbf{r}$ change. Consequently, $\mathbf{a}_1$ is tangent to the u_1-coordinate curve at P, and $\mathbf{a}_2$ is

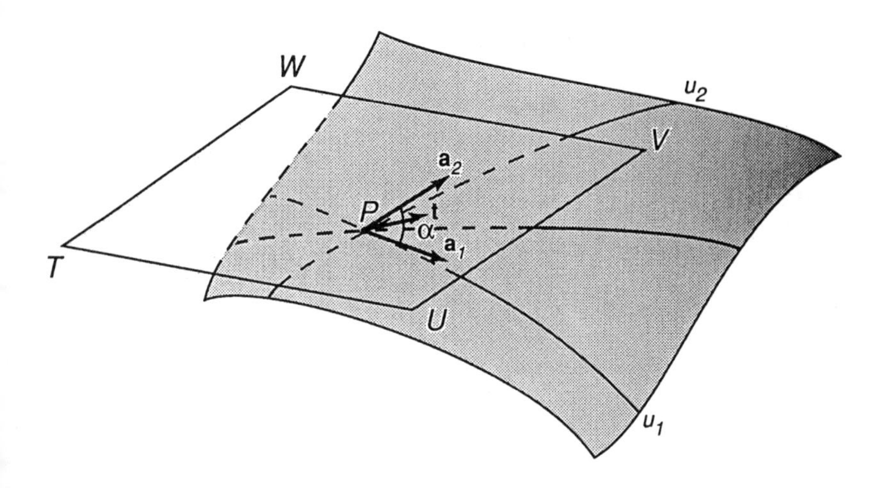

Figure 105 The vectors $\mathbf{a}_1, \mathbf{a}_2$, and $\mathbf{t}$ all lie in the tangent plane at P

tangent to the u_2-coordinate curve. The two vectors $\mathbf{a}_1, \mathbf{a}_2$ lie in the tangent plane at P (Fig. 105). Equation (12.46) shows how the unit tangent vector to PT can be written in terms of the tangent vectors to the two coordinate curves. As the increment $\Delta\mathbf{r}$ becomes infinitesimal, it falls into the tangent plane and is expressible by Equation (12.45).

Suppose now that we choose the arc PT in Fig. 104 to lie along the u_1-coordinate curve. For this case, $\Delta u_2 = 0$, $du_2 = 0$ in our previous calculations, and it is easy to see that the infinitesimal increment in $\mathbf{r}$ is

$$(d\mathbf{r})_1 = du_1\, \mathbf{a}_1 \,, \tag{12.47}$$

where the subscript 1 is attached to $d\mathbf{r}$ to emphasize that the increment refers to the u_1-coordinate curve. Likewise, choosing PT to lie along the u_2-coordinate curve, we have $\Delta u_1 = 0, du_1 = 0$, and

$$(d\mathbf{r})_2 = du_2\, \mathbf{a}_2 \,. \tag{12.48}$$

Let us denote the length of $(d\mathbf{r})_1$ by $(ds)_1$, and that of $(d\mathbf{r})_2$ by $(ds)_2$.

In view of Equations (12.47) and (12.48), we then have

$$(ds)_1^2 = du_1^2 \, ||\mathbf{a}_1||^2, \;\; (ds)_2^2 = du_2^2 \, ||\mathbf{a}_2||^2 \, , \tag{12.49}$$

where $||\mathbf{a}_1||$ and $||\mathbf{a}_2||$ the magnitudes of $\mathbf{a}_1$ and $\mathbf{a}_2$. These magnitudes can be calculated from Equations (12.44) using an expression of the type given by Equation (6.4). Comparing the results with Equations (12.26), and also recalling the discussion of the dot product in Chapter 6, we may deduce that

$$a_{11} = ||\mathbf{a}_1||^2 = \mathbf{a}_1 \cdot \mathbf{a}_1 \, ,$$

$$a_{22} = \;\; ||\mathbf{a}_2||^2 \;\; = \mathbf{a}_2 \cdot \mathbf{a}_2 \tag{12.50}$$

Thus, the coefficients a_{11} and a_{22} are the squares of the magnitudes of the tangent vectors to the surface-coordinate curves. Further, it is clear that

$$(ds)_1^2 = a_{11} \, du_1^2 \, , \;\; (ds)_2^2 = a_{22} \, du_2^2 \, . \tag{12.51}$$

Once surface coordinates are chosen, a_{11} and a_{22} can be obtained from surface measurements (see Experiment 41).

What about the metric coefficient a_{12} in Equation (12.27)? We can arrive at an interpretation of this as follows: The infinitesimal increment $d\mathbf{r}$ lies in the tangent plane at P and, as indicated in Equation (12.45), can be expressed as the vector sum of the two vectors $du_1 \, \mathbf{a}_1$ and $du_2 \, \mathbf{a}_2$. In the tangent plane at P, we therefore have a parallelogram $PA'C'B'$ of the type shown in Fig. 106. Let α be the measure of the angle between $\mathbf{a}_1$ and $\mathbf{a}_2$ ($0 < \alpha < \pi$). Using the dot product operation, we have

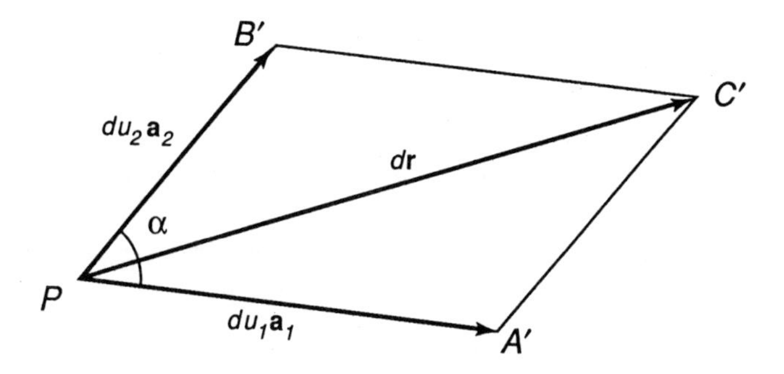

Figure 106 Infinitesimal parallelogram in tangent plane at P

$$
\begin{aligned}
ds^2 &= d\mathbf{r} \cdot d\mathbf{r} \\
&= (du_1\mathbf{a}_1 + du_2\mathbf{a}_2) \,.\, (du_1\mathbf{a}_1 + du_2\mathbf{a}_2) \\
&= du_1^2\, \mathbf{a}_1 \cdot \mathbf{a}_1 + 2du_1du_2\, \mathbf{a}_1 \cdot \mathbf{a}_2 + du_2^2\, \mathbf{a}_2 \cdot \mathbf{a}_2 \\
&= du_1^2\, ||\mathbf{a}_1||^2 + du_2^2\, ||\mathbf{a}_2||^2 + 2du_1du_2\, ||\mathbf{a}_1||\, ||\mathbf{a}_2||\, \cos\alpha \,.
\end{aligned}
\tag{12.52}
$$

With the aid of Equations (12.50), Equation (12.52) reduces to

$$
ds^2 = a_{11}\, du_1^2 + a_{22}\, du_2^2 + 2\, ||\mathbf{a}_1||\, ||\mathbf{a}_2||\, \cos\alpha\, du_1du_2 \,. \tag{12.53}
$$

But, ds^2 is also given by the first fundamental form in Equation (12.27). Subtracting Equation (12.27) from Equation (12.53), and utilizing Equations (12.50) once more, we obtain

$$
\cos\alpha = \frac{a_{12}}{\sqrt{a_{11}}\,\sqrt{a_{22}}} \,. \tag{12.54}
$$

Once a_{11} and a_{22} have been obtained from measurements, a_{12} can be found by measuring the angle α (see Experiment 41).

We also observe that $\mathbf{a}_{12}$ can be expressed in the form of a dot product (see Chapter 6):

$$a_{12} = \mathbf{a}_1 \cdot \mathbf{a}_2 \,. \tag{12.55}$$

Further, it is obvious that the surface coordinate lines are perpendicular to one another at P $(i.e., \alpha = \pi/2)$ if and only if $a_{12} = 0$.

Experiment 41 (Metric coefficients): (a) Mark off two rectangular axes, AB and AC, each about 2 inches long, on a sheet of squared paper and complete the figure as shown in Fig. 107. Cut out a narrow strip containing the line $DBACE$. (There is a horizontal cut at E.) Take a mixing bowl (or some other container) and carefully lay out the strip on it as follows: Tape the corner A to the bowl; run your finger along AC to make it lie snugly on the surface and tape it midway and at C; do the same for CE; likewise, tape AB, and finally BD to the bowl.

On the bowl, AB may be chosen as the u_1-coordinate line, and AC as the u_2-coordinate line. If we regard each marking as a coordinate unit, then since the lengths of the strips are not changed by the process of laying them on the surface, we see from Equations (12.51) that $\sqrt{a_{11}}$ is found simply by dividing the length of AB by the number of markings on it, and similarly for $\sqrt{a_{22}}$. Further, the 90°-angle at A will still measure 90° on the surface, and hence $a_{12} = 0$ at A.

Measure the angle that BD makes with CE. (To do this, you can use a flexible plastic protector, or alternatively, you can lay two narrow strips over BD and CE, tape the new strips to one another, lay them on a flat surface, and then measure the angle.) The coefficient a_{12} can then be calculated using Equation (12.54).

(b) Repeat the above procedure at various locations on the bowl. Record your observations and questions.

(c) Mark a rectangular Cartesian grid on a piece of rubber sheet, and stretch it over the surface of a bowl or vase. Measure the lengths of the surface-coordinate lines using a thread and a ruler. Also measure the angle between the coordinate lines at some points on the grid. Calculate a_{11}, a_{22}, a_{12}. □

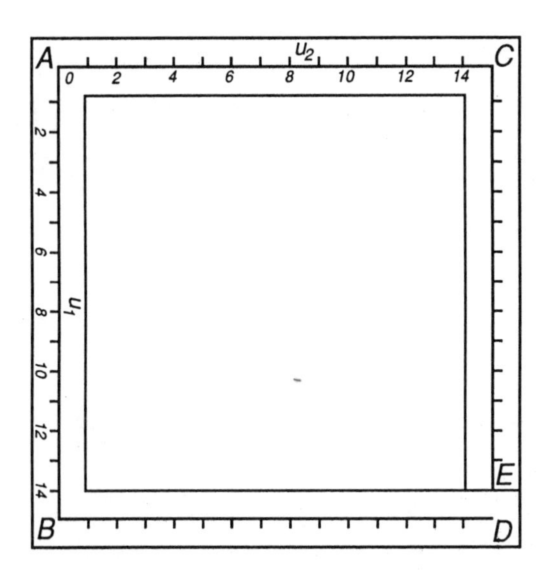

Figure 107 Preparing to lay out surface coordinates

Surface Area

In defining arc length in Chapter 8, we approximated the arc by a sequence of inscribed broken lines (Fig. 48). This provides a hint for defining surface area. Suppose that we choose many points on a piece of a surface, such as that shown in Fig. 108, and use them as vertices of plane triangles. As we choose more and more points, and construct smaller and smaller triangles, we might expect that, at least for sufficiently nice surfaces, the sum of the areas of the triangles would provide a better and better approximation to the area of the surface. But, there is a problem: even for such a nice surface as the cylinder, it is possible to construct sequences of inscribed triangles whose limits are very different from one another, and even some sequences whose areas go to infinity! This is called the *Schwarz paradox*.[3] The underlying reason is that some or all of the triangles may actually be falling away from the surface as their

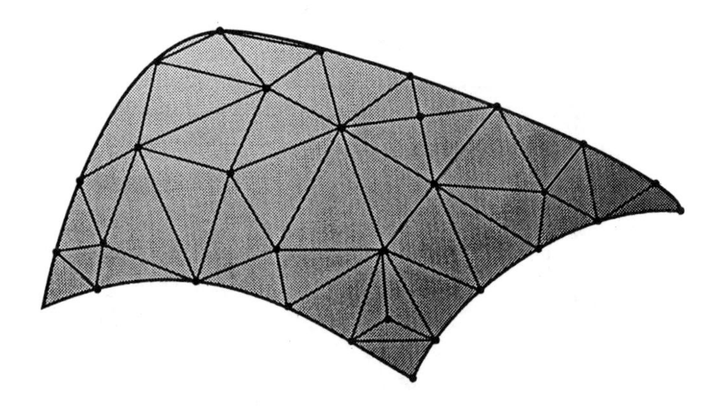

Figure 108 An array of triangles with all their vertices on a given surface

number increases. An illustration of how this can happen is provided in the next experiment.

Experiment 42 (The Schwarz paradox): Cut out two identical circles, each about 4 inches in diameter, from fairly heavy cardboard and inscribe squares $ABCD$ and $A'B'C'D'$ in them. Make small holes as close as you can to the vertices of the squares. Rotate one square by 45° relative to the other and glue one of the squares to the top of a match-box and the other to the bottom (Fig. 109). Use a thin string and a needle to construct the eight triangles $AA'B$, $BA'B'$, $B'BC$, $CB'C'$, $C'CD$, $DC'D'$, $D'DA$, $AD'A'$. You now have a set of triangles which you can regard as being inscribed on a short piece of a cylinder, of which the two circles are cross sections. (You can build a model with several sections if you like; also, polygons other than a square can be inscribed in the circles.) Examine how the string triangles are disposed relative to the tangent plane of the cylinder. How do they fall as the circles are pressed together? Do the areas of the triangles approach the area of the cylindrical segment as the height of the segment is decreased?

Suppose that you were to take a fixed vertical cylinder and use horizontal circles to divide it into m segments, and inscribe triangles in

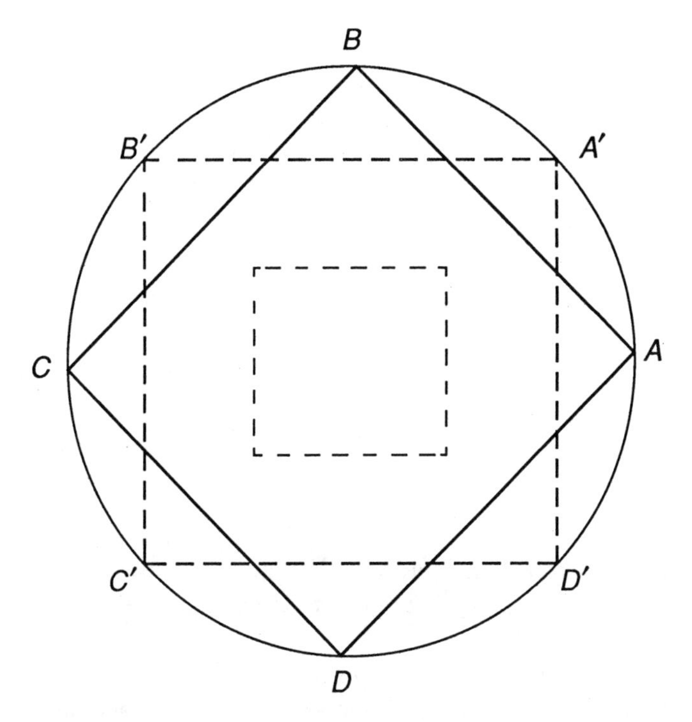

Figure 109 A model to illustrate the Schwarz paradox

each segment in the manner indicated above. What do you think would happen to the sum of the areas of the triangles as m goes to infinity? □

One way out of the dilemma is to insist that for a smooth surface each approximating triangle falls into the tangent plane to the surface as the size of the triangle decreases. An even better way is to use pieces of the tangent planes themselves to approximate the surface. Thus, referring back to Fig. 106, we may think of the infinitesimal parallelogram $PA'C'B'$ in the tangent plane at P as having the same area as the corresponding infinitesimal piece of surface in Fig. 105. Denoting this infinitesimal area by da and noting that the lengths of the sides of the parallelogram $PA'B'C'$ are $(ds)_1$ and $(ds)_2$, we have

$$\begin{aligned} da^2 &= [(ds)_1(ds)_2 \sin \alpha]^2 \\ &= (ds)_1^2(ds)_2^2(1 - \cos^2 \alpha) . \end{aligned} \tag{12.56}$$

Then, recalling Equations (12.51) and (12.54), we deduce that

$$da^2 = (a_{11}a_{22} - a_{12}^2)du_1^2du_2^2 , \tag{12.57}$$

and hence,

$$da = \sqrt{a_{11}a_{22} - a_{12}^2}\, du_1du_2 . \tag{12.58}$$

Thus, in the expression for infinitesimal area, the metric coefficients a_{11}, a_{12}, a_{22} play a fundamental role. The area of a smooth piece of a surface is given by the integral of da over the piece.

Example 5: A special spherical segment. We may calculate the area of a spherical segment S bounded by latitudes $\phi = 0$ and ϕ, and longitudes $\theta = 0$ and θ, as follows:

Since the metric coefficients for a sphere are expressed by Equations (12.35), Equation (12.58) reduces to

$$da = R^2 \cos \phi \; d\theta d\phi . \tag{12.59}$$

The area of the spherical segment is given by the integral

$$
\begin{aligned}
a = \int_S da &= \int_{\phi=0}^{\phi} \int_{\theta=0}^{\theta} R^2 \cos\phi \, d\theta d\phi \\
&= R^2 \int_{\phi=0}^{\phi} \cos\phi \left(\int_0^{\theta} d\theta\right) d\phi \\
&= R^2\theta \int_0^{\phi} \cos\phi \, d\phi \\
&= R^2\theta \sin\phi \, .
\end{aligned}
\tag{12.60}
$$

The area of a hemisphere can be found from Equation (12.56) by setting $\theta = 2\pi$ and $\phi = \pi/2$: $a = 2\pi R^2$. The area of a sphere is double this.

Notes to Chapter 12

[1] The formula (12.27) appears in Gauss's paper *General Investigations of Curved Surfaces*, which was published in Latin in 1827. An English translation is listed in the Bibliography, under Gauss (1827). Other important results contained in Gauss's paper will be discussed in this and the next few chapters. For modern treatments of differential geometry, the reader may consult Struik (1961), or Spivak (1970), or McCleary (1994), or any of several other excellent books available in a university library.

[2] It is an example of a *quadratic differential form*: *quadratic* because of the presence of second-degree terms, and *differential* because of du_1, du_2 ; the word *form* is used in its algebraic sense – it signifies a polynomial in which all terms are of the same degree (for example, $x^3 + xy^2 + x^2y$ is a form of degree 3 in two variables).

[3] H.A. Schwarz (1843-1921) was a German mathematician (see the *Dictionary of Scientific Biography* for an essay of his life and work). He made contributions to complex-function theory and the theory of minimal surfaces, and was renowned for his deep geometrical intuition.

13

Intrinsic Geometry of a Surface

The most deeply rooted geometrical properties of a surface are its topological ones: these are preserved under all homeomorphisms of the surface. Hence also, they are preserved under all deformations. (Some examples of deformations of surfaces were studied at the end of Chapter 11.) There is a broader class of properties that are intimately bound up with the geometry of the surface and that are preserved under a large subclass of homeomorphisms. These are ones that Gauss discovered, and which we will now explore.

Experiment 43 (Geometry on a cowl): Take a sheet of light cardboard, cut a slit in it, and at the end of the slit cut away a circular piece of material. Bend the sheet to form a hooded shape, or cowl (with a hole on top). [You can make more slits and holes to get more interesting shapes.] Staple the cardboard in place and trim. On a small rectangular piece of paper, construct a right-angled triangle. Measure the remaining angles and sides. Cut out the triangle and lay it on the surface of the cowl. Measure the sides and angles of the curvilinear triangle. What is the angle sum? Are the curvatures of the sides of the curvilinear triangle zero? Try other figures and measurements. Explain your findings. □

This experiment demonstrates that in some important ways, the geometry of figures on certain surfaces is the same as their geometry in the Euclidean plane. It shows that some features of surface geometry are independent of the shape of the surface in space (while some other features, of course, are not). Gauss was the first to realize that surfaces have a definite inner, or *intrinsic*, metric geometry that is independent of their shape in space (as long as they are not stretched or torn or made to overlap). If we take a surface made of flexible but inextensible material, its intrinsic geometry remains the same no matter what shape the surface is bent into. This is why the intrinsic geometry of a piece of paper is the same after the paper is bent into a cylindrical, or conical, or other possible shape. Geometrical properties that depend in any way upon the shape of the surface in the enveloping three-dimensional space are called *extrinsic*. The curvature of a surface curve is an example of an extrinsic property.

The intrinsic geometry of a surface can be determined from observations and measurements carried out on the surface itself, ignoring completely the manner in which the surface occupies the surrounding space. Appealing only to surface observations, we can build up a body of geometrical facts about figures drawn on the surface. The corresponding geometrical theory is even richer than Euclid's plane geometry, and can be very different from it.

Gauss recognized that the intrinsic metrical geometry of a smooth piece of surface is characterized by the expression for the length of the infinitesimal line element, *i.e.*, by

$$ds = \sqrt{a_{11}\, du_1^2 + 2a_{12}\, du_1 du_2 + a_{22}\, du_2^2}\,, \tag{13.1}$$

gotten From the first fundamental form in Equation (12.27). As we have seen, the metric coefficients a_{11}, a_{12}, a_{22} can be obtained from measurements made on the surface. In general, they vary across the surface.

If we inscribe Gaussian coordinates u_1, u_2 on a surface, we can find the coefficients a_{11}, a_{22}, a_{12} at a point by measuring two lengths and an angle as in Experiment 41. If we now bend the surface into another smooth shape without stretching any line in it, these coefficients will retain their original values. The length of an infinitesimal line element on the bent (smooth) surface will still be given by Equation (13.1).

It is obvious from the foregoing remark that the length of a surface curve is an intrinsic property, and hence surface-distance is also an intrinsic property. Since the area of an infinitesimal quadrilateral element is given by Equation (12.57), it is clear that surface area is an intrinsic property as well.

More generally, in addition to bending we could allow perfect cut-and-join operations to occur (see Experiment 8): these would not alter the intrinsic geometry of a surface. All topological properties of a surface are automatically intrinsic.

Experiment 44 (Intrinsic geometry of a sphere): Take a thin-walled plastic ball and cut it into two unequal pieces. Draw some identical figures on both pieces (*e.g.,* intersecting lines, triangles, a coordinate net). Bend one of the pieces and compare the figures on it to those on the other piece. Measure some intrinsic properties and discuss your findings. (Keep the two pieces for use in later experiments.) □

Many authors like to explain intrinsic geometry by referring to hypothetical intelligent two-dimensional beings that reside (and do mathematics!) on surfaces. They move around the surface, but cannot experience any geometrical object (such as the surface normal) that sticks out of the surface or lies completely off the surface. These surface-dwellers could study lines and figures on the surface. The only geometry that would be available to them experientially is the intrinsic geometry of the surface. Of course, it they had powerful enough imaginations, they might hypothesize the existence of three–and maybe more–geometrical dimensions.

Developability

In Experiment 43, you will have seen that certain familiar Euclidean results continue to hold on a bent surface. It is worthwhile to examine a little more closely the process that takes a plane surface into a bent one, and *vice versa.*

Let P be an arbitrary point on an elementary surface $\mathcal{S}$, such as the one illustrated in Fig. 83. Recall that the intersection $\mathcal{N}$ which $\mathcal{S}$ makes with some closed ball, of positive radius and centered at P, can be deformed into a closed disk $\mathcal{N}'$ in the plane. In general, lengths are altered during this deformation, and the intrinsic geometry of $\mathcal{S}$ is not the same as the geometry of a piece of the Euclidean plane. (Think of a sphere, for example.) However, for special surfaces, a length-preserving, or *isometric*, deformation that takes $\mathcal{N}$ into a disk in the plane may exist. In this case, we say that $\mathcal{S}$ is *developable at the point* P, or that it is *locally developable at* P. Thus, a cylinder is developable at every one of its points, whereas a sphere is not developable at any of its points.[1]

We shall say that an elementary surface is *globally developable* if and only if there exists an isometric homeomorphism that takes the whole surface into some subset of the Euclidean plane (possibly the whole plane). Global developability implies local developability.

Experiment 45 (Developability): Let us return to the surfaces we had in Experiments 28-32.

(a) Obviously, the surface $ABCD$ in Fig. 84 is developable at each of its points. How about the surfaces $EFGHIJ$, $KLMNOPQ$, and the surface bounded by the lines $RSTUVWR$ and $\overline{R}\,\overline{S}\,\overline{T}\,\overline{U}\,\overline{V}\,\overline{W}\,\overline{R}$? Can a surface be developable at each of its points and yet not be globally developable?

(b) Examine the developability of the surface of a cylindrical container at points on its lateral surface, points on the top and bottom surfaces, and points on the rim. (You could cut pieces from a cylindrical cardboard container and try to flatten them.)

(c) Is the Möbius strip locally developable? Is it globally developable?

(d) Discuss the local and global developability of the surface(s) you created in Experiment 32.

(e) Try to lay small squares of paper snugly on a ball, and convince yourself that the ball is not developable at any of its points. Conclude that there is no such thing as a perfect planar map of the earth's surface. □

In Experiment 45, you will have discovered that the surface $KLMNOPQ$ in Fig. 84 is locally developable everywhere except at its vertex K. In the case of the surface bounded by the lines $RSTUVWR$ and $\overline{R}\,\overline{S}\,\overline{T}\,\overline{U}\,\overline{V}\,\overline{W}\,\overline{R}$, we have local developability at all points, but the surface is not globally developable. Notice that it is not simply connected either. The surface of a box is locally developable at all points except the vertices, so it is not globally developable. The surface of a can is developable at all points that do not belong to the top or bottom rims. It is not globally developable. No part of a sphere is locally developable.

In an isometric deformation of a surface, no overlapping, tearing, or puncturing is allowed, and in addition, no curve can change its length. An isometric deformation can change the shape which a surface takes in space, but it cannot change the intrinsic geometry of the surface. Extrinsic properties can, of course, be changed by isometric deformations. In isometric homeomorphisms, perfect cut-and-join operations are admissible, but again the intrinsic geometry of the surface is re-created.

When a surface is locally developable at a point P, we can deform it in the vicinity of P isometrically into a piece of the Euclidean plane. Its local intrinsic geometry, which is preserved during this process, must therefore be the same as that of a piece of the plane. We may say that the geometry of the surface is locally Euclidean. It is clear from Experiment 45 that a surface may be locally Euclidean at all, some, or none of its points.

Note to Chapter 13

[1] The possibility of developing a surface upon a plane was considered by Euler (1707-1783) in a paper of 1772. Gauss explored this idea further, and also considered the possibility of developing a given surface upon a non-planar surface.

14

Gauss (1777-1855)

Mathematics, like music or literature or art, is an activity for which we human beings possess immeasurable stores of talent and passion. It is a highly intellectual activity, but it should not be regarded as an elitist one. Even those of us who have never created a song, or a story, or a piece of mathematics, can still experience much pleasure from playing or listening to music, or from reading a book or attending a play, or from doing a calculation or studying a proof. Furthermore, after an initial period of practice, many of us become quite accomplished at an activity, and continue to derive pleasure from it throughout their lives. Some others among us have sufficient talent to become professional musicians, writers, or mathematicians. And, scattered throughout history, there are those rare individuals whose genius leaves us in awe. Thus, in music, Bach (1685-1750), Mozart (1756-1791), and Beethoven (1770-1827) seem to possess almost superhuman powers. In literature, we have Shakespeare (1564-1616), Milton (1608-1674), Goethe (1749-1832), and several others. In mathematics, Archimedes (287-212 B.C.), Newton (1642-1727), and Gauss are ranked at the top, but magnificent contributions were also made by a large number of others. To take just a few examples: Both Euler (1707-1783) and Cauchy (1789-1857) produced voluminous new results in pure and applied mathematics; Galois, during his brief, tragic life (1811-1832) discovered the deep connection between algebraic equations and groups; Riemann (about whom much more will be said in Chapter 17) gave us his profound ideas on complex-variable theory and on differential geometry; and Georg Cantor (1845-1918) created the theory of sets and also taught us how to deal with transfinite numbers.

Much is to be gained from studying the lives and the original writings of the great mathematicians. As we learn about their material and social circumstances, their schooling and self-education, their aspirations and disappointments, and their successes and failures, we begin to understand the personal and social factors that can help or hinder the development of an individual's talent. We also come to appreciate the importance of such qualities as honesty and good judgement, and the necessity for passion and commitment, in the pursuit of mathematics. Furthermore, knowledge of the prevailing philosophical, scientific, and mathematical viewpoints often helps us to understand why particular mathematical questions are regarded as interesting, or even crucial. When one knows the original motivation for considering a class of problems, one normally finds that it is easier to understand why the associated theory is set up in what might otherwise seem to be a peculiar manner. As regards the original writings of the pioneers of mathematics, while it is usually advisable to begin with a modern treatment of the subject-matter, nevertheless, it is always worthwhile to go back to the sources. There, one can experience directly the brilliance and depth of truly outstanding minds. Often, one can also sense the intellectual excitement that is engendered by mathematical and scientific discovery.

Three mathematicians–Gauss, Riemann, and Levi-Civita–created the central ideas around which the present book revolves. It is only fitting that we devote some time to learning more about these men.

A great deal is known about Gauss's life.[1] Born on 30 April 1777 to poor, but respected parents in Brunswick (Braunschweig), Germany, Carl Friedrich Gauss was destined to be recognized, even in his own lifetime, as one of the greatest mathematicians that the world has ever seen. Gauss lived a long full life, made all the more precious by a childhood brush with death: while playing beside the canal next to his house, he fell into the water, but was rescued just before sinking.

As a small boy, Carl demonstrated a remarkable talent for mental arithmetic, and soon began to uncover patterns in numbers. A story which Gauss himself was fond of telling is worth repeating. At the age of ten, he attended his first course in arithmetic. Just after Gauss started the course, the teacher asked the pupils to find the sum of the first one

Carl Friedrich Gauss (1777-1855)

hundred numbers. It was the custom that the pupil who was the first to finish a problem would place his slate on the teacher's table; the second would lay his slate on top of this, and so on. The problem had hardly been stated, when Carl threw his tablet on the table. The other students, some of whom were a few years older than him, continued to calculate. At the end of the hour, the pile of slates was turned over. On top lay Gauss's slate, on which, to the teacher's astonishment, was written only the number 5050. Many of the other students gave wrong answers. Gauss explained that he had gotten the answer by observing that the sum $1 + 2 + \ldots + 100$ can be thought of as $(1 + 100) + (2 + 99) + \ldots + (50 + 51)$, which produces 50 pairs each of which adds up to 101.

Gauss's extraordinary ability was eventually brought to the attention of the Duke of Brunswick, who graciously provided the financial support which was needed to further the boy's education. Gauss progressed rapidly in all subjects, but excelled in mathematics and classical languages. He continued to experiment with numbers and discovered many of their properties through his calculations.

Gauss entered Göttingen University in 1795, and was initially drawn more towards philology than mathematics. However, a dramatic discovery, made by him on 30 March 1796, caused him to change his mind. While investigating a class of algebraic equations whose solution corresponds to the geometrical problem of dividing a circle into equal arcs, Gauss found a proof that the regular 17-sided polygon could be constructed with ruler and compass. This was the first advance on such questions in over 2000 years.

Gauss carefully studied the writings of the best mathematicians, including Archimedes, Newton, Euler, Lagrange, and Fermat. He was especially struck by Newton's creativity and method. He soon began to realize that he too could create mathematics and shape his arguments into beautiful chains of deductions. By 1797, he had begun work on his first masterpiece, *Disquisitiones arithmeticae* [*Investigations of Arithmetic*], in which he presented a treasure of new results in number theory in a thoroughly systematic and modern form. Almost two hundred years after its publication in 1801, the *Disquisitiones arithmeticae* still has an appearance of freshness.

To catch a glimpse of Gauss's approach to number theory, let us consider his notion of congruence, which is introduced at the very beginning of the book: Let a and b be integers, and let m be a positive integer. If m divides evenly into $a - b$, then a is said to be congruent to b with "modulus" m, and we write

$$a \equiv b \pmod{m} .$$

For example,

$$36 \equiv 24 \pmod{12} , \quad 36 \equiv 24 \pmod{3} ,$$

$$-7 = 5 \pmod{4} .$$

The abstract notion of congruence has some surprisingly practical applications. Thus, a clock reads hours mod 12, and a calendar reads days of the week mod 7.

A congruence may also involve an unknown. For instance,

$$x^2 \equiv 1 \pmod{8}$$

is a congruence which has solutions

$$x \equiv 1, 3, 5, 7 \pmod{8} .$$

Many interesting questions arise regarding the existence and properties of solutions to congruences of various types.

Gauss completed his studies at Göttingen in 1798, and subsequently obtained his Ph.D. degree from the University of Helmstedt in 1799. His dissertation contained a proof of the *fundamental theorem of algebra*: If $p(z)$ is a polynomial of degree greater than or equal to one whose coefficients are complex numbers, then the equation $p(z) = 0$ has at least one solution (which is generally complex). Gauss later gave three other proofs of the same theorem. He believed that new insights into the connections between mathematical phenomena could be gained by trying to prove known results in new ways. He himself furnished eight proofs of one of his most famous theorems.

During these years, Gauss made many other mathematical discoveries, several of which remained unpublished until later in his life, and some of which only came to light after his death. He was not willing to publish any piece of work unless he was satisfied that he had treated it thoroughly, deeply, and clearly. He succeeded in emulating his heroes Archimedes and Newton in regard to the clarity and depth of his mathematical thought.

On New Year's Day of 1801, an event took place in the skies over Palermo that was to have an important influence on Gauss's career. This was the discovery and subsequent loss of a new planet, Ceres, by the Sicilian astronomer Piazzi.[2] Theoretical astronomers attempted to predict the orbit of Ceres from the sparse data of Piazzi for January and early February of 1801. Gauss entered the fray in the autumn of 1801. He calculated the elliptical orbit that would best match Piazzi's data. His predictions were published in early December 1801, and differed significantly from those of other skilled mathematical astronomers. Observations in December were marred by poor weather conditions, but on 1 January 1802, Ceres was found exactly where Gauss had predicted. This was an amazing feat, since the entire ellipse had to be calculated from data which covered only a small part of its circumference. For his calculations on Ceres, Gauss became famous throughout the scientific world. In March 1802 an even smaller planet, Pallas, was discovered, and again Gauss computed its orbit accurately.

These successes led Gauss to choose astronomy as a profession. He became director of the observatory at Göttingen University in 1807. Starting out in the old observatory in 1816, Gauss devoted enormous effort to procuring astronomical instruments of the highest available quality. But, he also found some time to produce new mathematical results. It was during these years that Gauss reflected upon the possibility of non-Euclidean geometry, but he was reluctant to publicize his views on such a controversial subject.

In 1805, Gauss married a young woman, Johanna Osthoff, whom he loved deeply. But, just four years into their blissful marriage, Johanna died soon after giving birth to the couple's third child. Gauss was grief-stricken. He did marry again, however, and had three more children. Unfortunately, the health of Gauss's second wife, Minna Waldeck, was frail and she lived only until 1831.

About 1818, Gauss undertook the arduous task of conducting an accurate land-survey of the kingdom of Hanover. This placed a serious burden on his time for several years. Still, he took pleasure in performing the enormous amount of computation that was involved,[3] in designing improved instrumentation, and in obtaining accurate measurements. As with many other fields that he touched, Gauss brought both theoretical and practical improvements to geodesy. And, most happily, it was as a result of his extended involvement with geodesy that a deep feeling for curved surfaces grew within Gauss. This culminated in his second masterpiece, *Disquisitiones generales circa superficies curvas* [*General Investigations of Curved Surfaces*], which he presented to the Royal Society of Göttingen on 8 October 1827. It was in this memoir that Gauss expounded his seminal ideas on the intrinsic geometry of surfaces. (Some of these ideas were discussed in Chapters 12 and 13, and the crowning concept among them, namely that of intrinsic curvature, will be studied in Chapter 16.)

The year 1828 was another turning point in Gauss's life. Having exhibited some signs of intellectual restlessness in the preceding few years, Gauss moved towards physics as a focus of concentration. He had had an interest in magnetism since 1803, and had been encouraged by his scientific friends, especially Alexander von Humboldt, to apply his mathematical skills to this emerging branch of physics. In 1828, at Humboldt's persuasion, Gauss went to Berlin to attend the only scientific convention of his entire life. He stayed at Humboldt's residence for three weeks and availed himself of his host's collection of magnetic instruments. He also met and was impressed by the talents of the young experimental physicist, Wilhelm Weber (1804-1891).

In 1830, the chair of physics at Göttingen University became vacant. Weber was appointed to this position in 1831, and an intimate friendship and fruitful collaboration with Gauss ensued: experiments were conducted on terrestrial magnetism and electromagnetism, theoretical ideas were developed, an iron-free magnetic observatory was built, a Magnetic Association was founded, and an electric telegraph was invented. The great Scottish physicist James Clerk Maxwell (1831-1879) had high praise for Gauss's work on magnetism:[4]

Gauss, as a member of the German Magnetic Union, brought his powerful intellect to bear on the theory of magnetism, and on the methods of observing it, and he not only added greatly to our knowledge of the theory of attractions, but reconstructed the whole of magnetic science as regards the instruments used, the methods of observation, and the calculation of the results, so that his memoirs on Terrestrial Magnetism may be taken as models of physical research by all those who are engaged in the measurement of any of the forces in nature.

Stimulated by the work of Oersted (1777-1851), Ampère (1775-1836), and Faraday (1791-1867), Gauss strove to develop an understanding of electromagnetism. In 1835, he discovered (but did not publish) a law of electrical attraction. Gauss was unable to satisfy himself regarding the conceptual foundations of electromagnetism. Instead, the honor of elucidating the theory was reserved for Maxwell who, basing his mathematics upon the physical ideas of Faraday, presented a revolutionary "field theory" of electromagneticism.

A political event severed the collaboration between Gauss and Weber. For some time, Hanover had been united with the British Empire, and had enjoyed a liberal constitution since 1833. The union was dissolved in 1837: Victoria became Queen of England and her uncle, Ernst August, acceded to the throne of Hanover. One of the king's first acts was to revoke the hard-won constitution of 1833. Weber, along with six other Göttingen professors (including the famous Grimm brothers) signed a letter of protest. They were all dismissed from their positions.

Despite Gauss's efforts on his behalf, Weber was unable to regain his position at that time and eventually moved to the University of Leipzig. However, in 1848, he was recalled to Göttingen. His friendship with Gauss continued thereafter, but on account of Gauss's advanced age, the magnificent collaboration of earlier years did not resume.

Towards the end of his life, Gauss was fortunate to have exceptionally good students at Göttingen. Foremost among these were Riemann (1826-

1866) and Dedekind (1831-1916), both of whom were to contribute profoundly to modern mathematics.

By early 1854, Gauss's health had deteriorated as a result of a heart condition. However, some improvement took place in the spring and summer, and he had the immense satisfaction of being present in June 1854 at the reading of Riemann's epoch-making paper on the foundations of geometry.

With the coming of fall, Gauss's health again worsened, and after a difficult period of suffering, his life came to an end in the early hours of 23 February 1855. He is buried in St. Albans Cemetery in Göttingen. In 1865, the king of Hanover, George V, dedicated a copper plaque over the door of Gauss's study. On it is written:[5]

> *. . . From here his immortal spirit ascended to heaven, in order to contemplate pure truth there in eternal light, whose mysterious doctrines he strove with holy seriousness to decipher here below from the starry writing of the firmament . . .*

Notes to Chapter 14

[1] I have benefitted from reading the biography of Gauss by G. Waldo Dunnington (listed in the Bibliography), as well as the memorial by Gauss's friend and colleague, Sartorius von Waltershausen (which was translated into English by Gauss's great-granddaughter), and the excellent account of Gauss's life and work given by K.O. May in *The Dictionary of Scientific Biography*. Dunnington's book contains several interesting portraits and photographs. Bell (1937), Muir (1961), Hollingdale (1989), and Simmons (1992) all have interesting chapters on Gauss.

[2] Ceres, measuring about 1000 kilometers in diameter, is the biggest member of a large class of rock-like objects called asteroids, which orbit the sun. The asteroids lie mainly in a band between the orbits of the Mars and Jupiter.

[3] Gauss estimated that he had worked on a million numbers in the survey.

[4] This excerpt is taken from the preface to the first edition of Maxwell's *Treatise on Electricity and Magnetism* (Maxwell, 1873).

[5] This is quoted from Dunnington (1955), p. 323.

15

Normal Sections

One way of exploring curved surfaces is to examine the curvature of curves lying on them. In the present chapter, we will use planes to cut a surface in a specific way, and we will study the resulting curves to gain some insight into the nature of surface curvature.

Bristling with Normals

Consider a point P on the side of a mixing bowl (Fig. 110), and let $TUVW$ be the tangent plane to the bowl at P. Let us attach a unit vector $\mathbf{N}$ at P that is perpendicular to the tangent plane and that points from the interior of the bowl towards the exterior. Performing the same

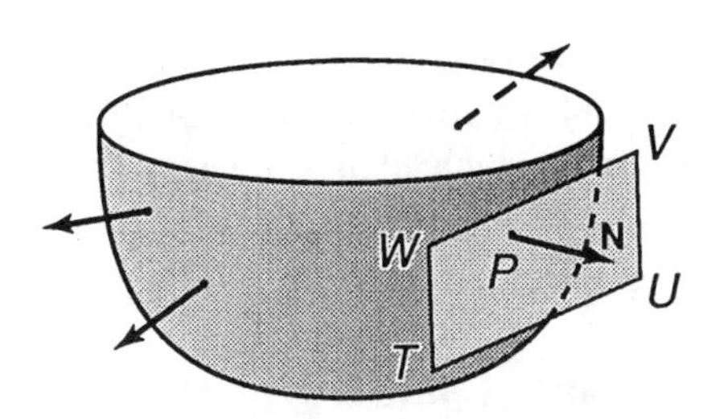

Figure 110 Outward unit normal vectors (or bristles) on a bowl

construction at all other points of the bowl except around the rim, we obtain the *outward unit normal* vector field to the bowl. For these vectors, we might also use the less formal but more evocative name of "bristles" (as on hedgehogs and toothbrushes!). The bristles on a sphere are easy to imagine, and so are those on a torus, and those on the sides of a cube.

Experiment 46 (Bristling a surface): Here are some ways of erecting normals to physical surfaces:

(a) Take an eggplant or watermelon, or a large apple or potato, and several toothpicks. Insert the toothpicks at several places on the surface. (You can gauge perpendicularity by using a right-angled triangle, or better yet, a right-angled triangle held at a right angle to a protractor.) Study how the directions of the normals change as you traverse the surface.

(b) Take a cardboard cylinder or cone, or the cowl in Experiment 43, and several of thumbtacks. Push the thumbtacks through from the interior. Secure them with tape. Watch how the bristles change their direction in space as you bend the cardboard. Are bristles part of the intrinsic geometry of a surface?

(c) You can use a light cardboard strip bristled with thumbtacks to study how the normal vector field varies across a surface.

(d) You can use a large rubber band bristled with thumbtacks to do the same thing. □

Orientability

For bowls, spheres, and watermelons, we can also easily imagine an *inward* unit normal field, in which all the outward unit normal vectors are reversed in sense. For some surfaces, however, distinguishing outward from inward becomes problematical, as the following experiment shows.

Experiment 47 (Inward or outward?): Take two fairly long strips of light cardboard (I cut 1-inch wide strips from a manila folder), several thumbtacks, and tape. Form one strip into a cylindrical band and the other into a Möbius band (see Experiment 30). Lay the strips side by side. Push a thumbtack through any point on the centerline of the cylindrical strip and push another thumbtack, aligned in the same way as the first one, through a point on the centerline of the Möbius strip. Push another thumbtack through the cylindrical strip about 1 inch from the first one. Do the same on the Möbius strip. Secure the thumbtacks with tape.

Continue this process around both strips until you arrive back at the first pair of thumbtacks. What do you find? □

Experiment 47 illustrates the fact that, for a Möbius strip, it is not possible to choose a unit normal vector field that varies continuously around the strip: after traversing the strip, the normal vector has switched its sense. Such a surface is said to be *non-orientable*. The plane, the sphere, the torus, and other surfaces for which a continuous normal vector-field can be assigned, are orientable. Locally, every smooth elementary surface is orientable.

Normal Sections

If you slice a small piece from an apple in the usual way, the plane of the slice will not contain the unit normal vectors to the apple at the boundary of the slice (Fig. 111 (a)). But, if you slice a cucumber in the usual way, the slice-plane contains the normals (Fig. 111 (b)). Consider a smooth orientable surface S (Fig. 112).[1] Let **N** be the outward or inward unit normal vector at a point P of S. The tangent plane at P is $TUVW$. Construct another plane $DEFG$ which passes through P and

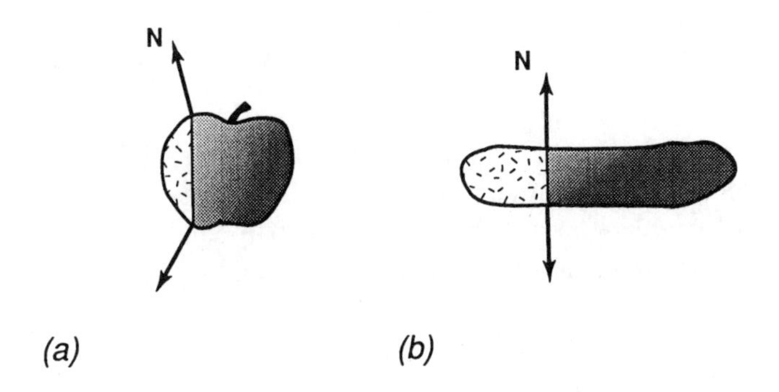

Figure 111 Slicing an apple and a cucumber

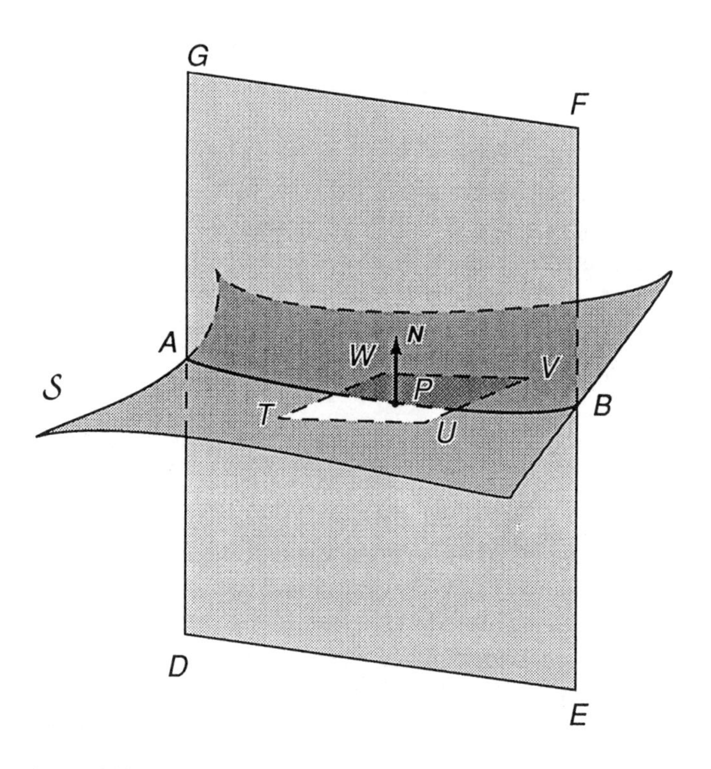

Figure 112 A normal section AB of a surface

is perpendicular to $TUVW$. The plane $DEFG$ therefore contains **N**. The intersection AB of $DEFG$ with the surface $\mathcal{S}$ is called a *normal section of* $\mathcal{S}$ *at* P.

Experiment 48 (Normal and non-normal sections): (a) Use toothpicks to identify the outward unit normal at a few points on a cucumber, large potato, eggplant, or watermelon. Use a sharp knife (be very careful!) to obtain normal and non-normal sections. Study the curves formed by the boundaries of the slices.

(b) Find a fairly stiff cardboard roll, or cylindrical cardboard container, and carefully cut normal and non-normal sections from it. Study the curves formed by the boundaries. Keep the objects that you used

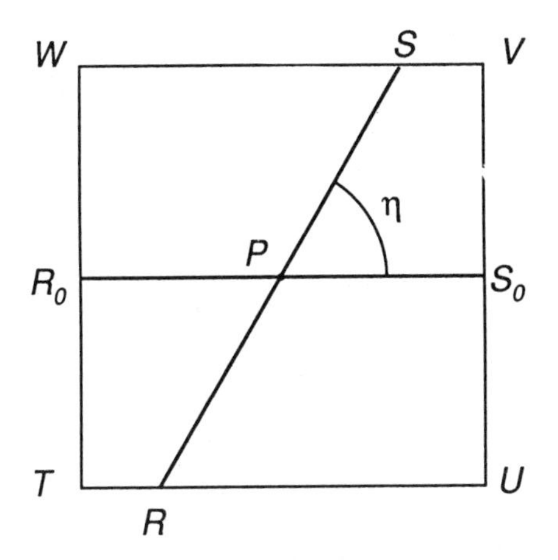

Figure 113 Parametrizing the normal sections at P

here for re-use in Experiments 49, 50 and 51. □

Referring again to Fig. 112, imagine that the sectioning plane $DEFG$ can be revolved as if it were supported on an axle that passes through P and is parallel to **N**. As $DEFG$ is rotated from one position into another, a different normal section of $\mathcal{S}$ is cut out at P. Correspondingly, a family of straight lines is cut out on the tangent plane. We can use these lines to obtain a parametrization of the normal sections at P. Thus, suppose that the plane $DEFG$ in Fig. 112 intersects the tangent plane $TUVW$ in RS. ($DEFG$ intersects $\mathcal{S}$ in AB; projecting AB perpendicularly onto $TUVW$ yields RS). We may choose any configuration of the revolving plane $DEFG$ as a fixed reference: let the chosen reference plane intersect $TUVW$ in R_0S_0 (Fig. 113). The angle η that RS makes with R_0S_0 can be used as a parameter. We call η the sectioning angle.

Experiment 49 (Studying normal sections): Choose three or more eggplants (or other firm, smooth-skinned vegetables or fruit) that are as identical as possible. Pick a point P having the same location on all of the eggplants and indicate the normal vectors with toothpicks. Mark off a reference line R_0S_0 on all of the eggplants. Choose different normal sections on each one, and slice along these. Compare the boundary curves of the slices with one another. Draw some conclusions about how these curves vary with the sectioning angle η. Save your material for Experiment 50. □

Curvature of Normal Sections

Consider again the normal section AB of S in Fig. 112. Let **t** be the unit tangent vector to AB at P (Fig. 114); **t** lies in the tangent plane $TUVW$, and hence is perpendicular to the surface normal **N**. Choose Gaussian coordinates u_1, u_2 on S, and let $\mathbf{a}_1$ and $\mathbf{a}_2$ be the tangent vectors to the coordinate curves. Denoting the arc length of AB by s, recall from Equation (12.46) that

$$\mathbf{t} = \frac{d\mathbf{r}}{ds} = \mathbf{a}_1 \frac{du_1}{ds} + \mathbf{a}_2 \frac{du_2}{ds} \quad . \tag{15.1}$$

Also note that since $\mathbf{a}_1$ and $\mathbf{a}_2$ lie in the tangent plane $TUVW$, we must have

$$\mathbf{N} \cdot \mathbf{a}_1 = 0, \ \mathbf{N} \cdot \mathbf{a}_2 = 0 \tag{15.2}$$

at P.

In Chapter 10, we saw that the curvature of a curve is given by Equation (10.12), in which **n** stands for the principal unit normal vector to the curve. We may apply this equation to calculate the curvature of AB in Fig. 114. Observe that, in the present case, both **n** and **N** lie in the plane *DEFG*, and both of them are perpendicular to **t**. Hence, **n** and **N** are parallel to one another, and we may in fact choose **n** to have the same sense as **N**, as indicated in Fig. 114. We denote the curvature of AB at P by κ_N and call it the *curvature of a normal section*. We thus have

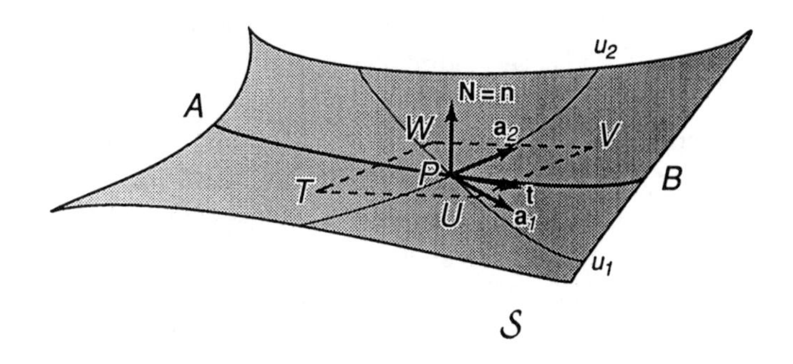

Figure 114 What is the curvature of the normal section AB?

$$\kappa_N \mathbf{N} = \frac{d\mathbf{t}}{ds} . \tag{15.3}$$

Taking a dot product of both sides of Equation (15.3) with **N**, and remembering that **N** is a unit vector, we obtain

$$\kappa_N = \frac{d\mathbf{t}}{ds} \cdot \mathbf{N} . \tag{15.4}$$

How does κ_N vary as we rotate the sectioning plane about **N** to produce different normal sections? Let us take some examples. For the trivial case of a plane, each normal section through any given point intersects the plane in a line, and hence, $\kappa_N = 0$ always. At any point P on a sphere of radius R, each normal section produces an arc of a great circle, and so, with the surface normal **N** pointing inwards, we have $\kappa_N = 1/R$ for all sections through P (see Equation (10.4)). At a point on a cylinder, the normal section varies from being a circle (transverse cut) to being a straight line (longitudinal cut); correspondingly, κ_N varies from (1/ radius) to 0. On a saddle-shaped surface, such as the region of your hand between the thumb and index finger, or a mountain pass, κ_N takes on both negative and positive values.

If every normal section through P has the same curvature, P is said to be an *umbilic* (or *navel*). Thus, every point on a sphere is an umbilic, and more trivially, so is every point on a plane.

Experiment 50 (Curvature of normal sections): (a) Take a cylindrical roll, or container, of medium-weight cardboard and cut a normal section across it. (I used a container of thickness 1 mm and radius 75 mm.) Trace the normal section on a piece of paper and estimate its curvature (see Experiment 23).

(b) Do the same for a normal section cut from a large potato or an eggplant.

(c) On various surfaces, try to find normal sections for which $\kappa_N = 0$. How many such sections do you find for a point on (i) a cylinder; (ii) an eggplant; (iii) a vase? How about at various points on a bicycle tube? □

We now proceed to derive an important analytical expression for κ_N. Thus, starting out with Equation (15.1) and taking the derivative of $\mathbf{t}$ with respect to s, we find that

$$\frac{d\mathbf{t}}{ds} = \frac{d\mathbf{a}_1}{ds}\frac{du_1}{ds} + \frac{d\mathbf{a}_2}{ds}\frac{du_2}{ds} + \mathbf{a}_1\frac{d^2u_1}{ds^2} + \mathbf{a}_2\frac{d^2u_2}{ds} \tag{15.5}$$

where the product rule for differentiation has been used. Taking a dot product of both sides of Equation (15.5) with $\mathbf{N}$, and recalling Equations (15.2) and (15.4), we deduce that

$$\kappa_N = \left(\frac{d\mathbf{a}_1}{ds}\frac{du_1}{ds} + \frac{d\mathbf{a}_2}{ds}\frac{du_2}{ds}\right) \cdot \mathbf{N}\,. \tag{15.6}$$

Next, employing the chain rule of calculus, we find that

$$\begin{aligned}\frac{d\mathbf{a}_1}{ds} &= \frac{\partial\mathbf{a}_1}{\partial u_1}\frac{du_1}{ds} + \frac{\partial\mathbf{a}_1}{\partial u_2}\frac{du_2}{ds} \\ &= \frac{\partial^2\hat{\mathbf{r}}}{\partial u_1^2}\frac{du_1}{ds} + \frac{\partial^2\hat{\mathbf{r}}}{\partial u_1\partial u_2}\frac{du_2}{ds}\,,\end{aligned}$$

$$\frac{d\mathbf{a}_2}{ds} = \frac{\partial \mathbf{a}_2}{\partial u_1}\frac{du_1}{ds} + \frac{\partial \mathbf{a}_2}{\partial u_2}\frac{du_2}{ds}$$

$$= \frac{\partial^2 \hat{\mathbf{r}}}{\partial u_2 \partial u_1}\frac{du_1}{ds} + \frac{\partial^2 \hat{\mathbf{r}}}{\partial u_2^2}\frac{du_2}{ds}\,, \tag{15.7}$$

where use has also been made of Equations (12.44). We now introduce the abbreviations

$$b_{11} = \frac{\partial^2 \hat{\mathbf{r}}}{\partial u_1^2} \cdot \mathbf{N},\;\; b_{22} = \frac{\partial^2 \hat{\mathbf{r}}}{\partial u_2^2} \cdot \mathbf{N}$$

$$b_{12} = \frac{\partial^2 \hat{\mathbf{r}}}{\partial u_1 \partial u_2} \cdot \mathbf{N} = \frac{\partial^2 \hat{\mathbf{r}}}{\partial u_2 \partial u_1} \cdot \mathbf{N} = b_{21}\;. \tag{15.8}$$

From Equations (15.7) and (15.8), it follows that

$$\frac{d\mathbf{a}_1}{ds} \cdot \mathbf{N} = b_{11}\frac{du_1}{ds} + b_{12}\frac{du_2}{ds}\,,$$

$$\frac{d\mathbf{a}_2}{ds} \cdot \mathbf{N} = b_{21}\frac{du_1}{ds} + b_{22}\frac{du_2}{ds}\,. \tag{15.9}$$

Consequently, Equation (15.6) becomes

$$\kappa_N = b_{11}\left(\frac{du_1}{ds}\right)^2 + 2b_{12}\frac{du_1}{ds}\frac{du_2}{ds} + b_{22}\left(\frac{du_2}{ds}\right)^2\,. \tag{15.10}$$

Writing Equation (15.10) as a differential form, we have

$$\kappa_N ds^2 = b_{11}du_1^2 + 2b_{12}du_1du_2 + b_{22}du_2^2\,. \tag{15.11}$$

The quadratic form on the right-hand side of Equation (15.11) is called the *second fundamental form.* It has the same algebraic structure as the first fundamental form, which we met in Equation (12.27). Note, however, that the coefficients b_{11}, b_{22}, b_{12} involve the second partial derivatives of the position-vector function $\hat{\mathbf{r}}$ of the surface, whereas the coefficients a_{11}, a_{22}, a_{12} involve the first partial derivatives of this function.

Substituting Equation (12.27) in Equation (15.11), we find that

$$\kappa_N = \frac{b_{11}du_1^2 + 2b_{12}du_1du_2 + b_{22}du_2^2}{a_{11}du_1^2 + 2a_{12}du_1du_2 + a_{22}du_2^2} . \tag{15.12}$$

In other words, the curvature of a normal at P section is given by the quotient of the second and first fundamental forms. Observe that since $ds^2 > 0$, the sign of κ_N is always the same as the sign of the second fundamental form.

If we choose a normal section at P such that the tangent vector $\mathbf{t}$ lies along the u_1-coordinate curve, then $du_2 = 0$ and the corresponding normal curvature is

$$\kappa_N = \frac{b_{11}}{a_{11}} \quad (du_2 = 0) . \tag{15.13a}$$

Similarly, for a normal section tangent to the u_2-coordinate curve, we have

$$\kappa_N = \frac{b_{22}}{a_{22}} \quad (du_1 = 0) . \tag{15.13b}$$

The Principal Curvatures

For a given normal section AB through the point P in Fig. 114, the curvature κ_N of AB can be found by substituting the components $du_1/ds, du_2/ds$ of the tangent vector $\mathbf{t}$ (see Equation (15.1)) in Equation

(15.10) In general, different normal sections at P will have different curvatures. But, does κ_N vary in any special way as we rotate the sectioning plane about the normal $\mathbf{N}$?

To pursue this question analytically, let us parametrize the normal sections by the sectioning angle η (Fig. 113). Both $\mathbf{t}$ and κ_N then became functions of η. We are interested in how the function $\kappa_N(\eta)$ behaves for values of η in the closed interval $0 \leq \eta \leq \pi$ radians.

We may simplify our calculations by choosing Gaussian coordinates such that, at the point P, $\mathbf{a}_1$ and $\mathbf{a}_2$ are unit vectors $\mathbf{e}_1$ and $\mathbf{e}_2$ and also that these are perpendicular to one another. Of course, even a small distance from P, this may no longer be true (see Experiment 41). But, in the present calculation, we will make use of the values of $\mathbf{a}_1$ and $\mathbf{a}_2$ only at P, and what happens to them at other points will not affect our results.

In Fig. 115, the tangent plane $TUVW$ at P is drawn. On it are shown its intersections RS and R_0S_0 with the the sectioning plane $DEFG$ and with a reference configuration of the latter, respectively (see Fig. 113). The unit tangent vectors $\mathbf{t}$ can now be conveniently expressed as

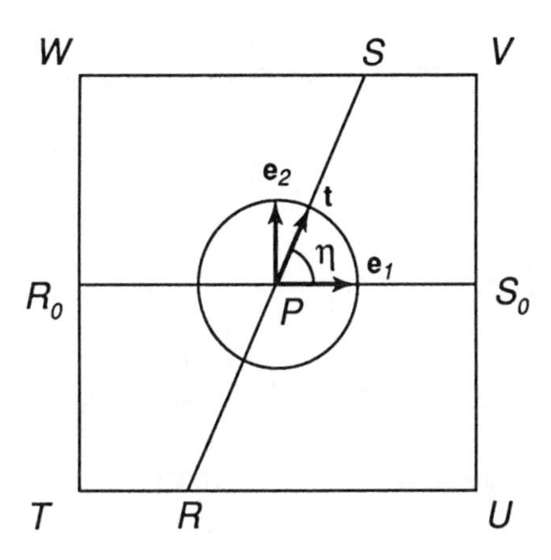

Figure 115 Setting up a pair of perpendicular unit vectors $\mathbf{e}_1$, $\mathbf{e}_2$ at P

$$\mathbf{t} = \cos\eta\, \mathbf{e}_1 + \sin\eta\, \mathbf{e}_2 \,. \tag{15.14}$$

Substituting the components of **t** from this equation in Equation (15.10), we deduce that

$$\kappa_N = \kappa_N(\eta) = b_{11}\cos^2\eta + 2b_{12}\sin\eta\cos\eta + b_{22}\sin^2\eta \,. \tag{15.15}$$

Thus, at the point P, the coefficients b_{11}, b_{12}, b_{22} have definite numerical values, and as we vary η, the formula (15.15) delivers the appropriate value of κ_N. In particular, the curvature of the reference normal section is found by setting $\eta = 0$ in (15.15):

$$\kappa_N(0) = b_{11} \,. \tag{15.16a}$$

Likewise, the curvature of the normal section containing $\mathbf{e}_2$ is

$$\kappa_N\left(\frac{\pi}{2}\right) = b_{22} \,. \tag{15.16b}$$

The curvature of the normal section lying midway between $\mathbf{e}_1$ and $\mathbf{e}_2$ is

$$\kappa_N\left(\frac{\pi}{4}\right) = \frac{b_{11} + b_{22}}{2} + b_{12} \,. \tag{15.16c}$$

Using the trigonometric identities

$$\cos 2\eta = 2\cos^2\eta - 1,\ \sin 2\eta = 2\sin\eta\cos\eta \,, \tag{15.17}$$

we may rewrite Equation (15.15) in the form

$$\kappa_N(\eta) = \frac{b_{11} + b_{22}}{2} + \frac{b_{11} - b_{22}}{2}\cos 2\eta + b_{12}\sin 2\eta \,. \tag{15.18}$$

This function will attain a maximum at some value of η $(0 \leq \eta \leq \pi)$, and will also attain a minimum at some value of η in the same interval; furthermore, it will take on all values in between the maximum and the minimum. (In some important special cases, the minimum and maximum values coincide.) We may search for the maximum and minimum values of $\kappa_N(\eta)$ by setting the derivative $d\kappa_N/d\eta$ to zero. Thus, differentiating the right-hand side of Equation (15.18), we have

$$0 = (\frac{b_{11} - b_{22}}{2})(-2 \sin 2\eta) + 2b_{12} \cos 2\eta \tag{15.19a}$$

or equivalently,

$$\frac{b_{11} - b_{22}}{2} \sin 2\eta = b_{12} \cos 2\eta \,. \tag{15.19b}$$

For given values of b_{11}, b_{12}, b_{22}, we want to solve Equation (15.19b) for η. Let us consider the two cases $b_{11} = b_{22}$ and $b_{11} \neq b_{22}$ separately.

Case 1 ($b_{11} = b_{22}$). In this case, Equation (15.15) reduces to

$$\kappa_N(\eta) = b_{11} + b_{12} \sin 2\eta \,. \tag{15.20}$$

The maximum value of $\sin 2\eta$ is 1 and occurs at $2\eta = \pi/2$, and its minimum is -1, occurring at $2\eta = 3\pi/2$. If b_{12} is also zero, $\kappa_N(\eta)$ is constant, and hence the point P is an umbilic. If b_{12} is positive, the maximum value of $\kappa_N(\eta)$ is $b_{11} + b_{12}$ and occurs at $\eta = \pi/4$, while the minimum value is $b_{11} - b_{12}$ and occurs at $\eta = 3\pi/4$. If b_{12} is negative, the maximum is $b_{11} - b_{12}$ and the minimum is $b_{11} + b_{12}$, occurring at $\eta = 3\pi/4$ and $\eta = \pi/4$, respectively.

Case 2 ($b_{11} \neq b_{22}$). In this case, Equation (15.19b) can be expressed as

$$\tan 2\eta = \frac{b_{12}}{\frac{1}{2}(b_{11} - b_{22})} \,. \tag{15.21}$$

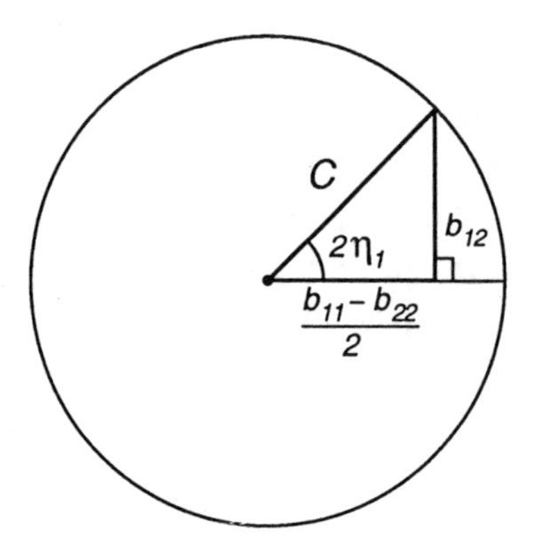

Figure 116 Representation of solution for angle $2\eta_1$

Examination of the graph of the tangent function reveals that at each value of the right-hand side of Equation (15.21), there will be two values of the angle 2η that produce it: one will occur in the interval $0 \leq 2\eta < \pi$ and the other in the interval $\pi \leq \eta < 2\pi$, and they will be π radians apart. Recalling the definitions of the trigonometric functions, we see that if we draw a circle (Fig. 116) of radius

$$C = \sqrt{b_{12}^2 + \left(\frac{b_{11} - b_{22}}{2}\right)^2}, \tag{15.22}$$

then for one of the solutions (call it η_1), the values of $\sin 2\eta_1$ and $\cos 2\eta_1$ are given by

$$\begin{aligned} C \sin 2\eta_1 &= b_{12}, \\ C \cos 2\eta_1 &= \frac{b_{11} - b_{22}}{2}. \end{aligned} \tag{15.23}$$

For the other solution, since $2\eta_2 = 2\eta_1 \pm \pi$, we will have

$$C \sin 2\eta_2 = C \sin(2\eta_1 \pm \pi) = -C \sin 2\eta_1 = -b_{12}\,, \qquad (15.24)$$

$$C \cos 2\eta_2 = C \cos(2\eta_1 \pm \pi) = -C \cos 2\eta_1 = -\frac{b_{11} - b_{22}}{2}\,.$$

Let us denote the values of κ_N corresponding to the solutions η_1 and η_2 by κ_1 and κ_2, respectively. Then, utilizing Equations (15.23) in Equation (15.18), we find that

$$\kappa_1 = \frac{b_{11} + b_{22}}{2} + C\,. \qquad (15.25a)$$

Likewise, utilizing Equations (15.24) in Equation (15.18), we obtain

$$\kappa_2 = \frac{b_{11} + b_{22}}{2} - C\,. \qquad (15.25b)$$

Consequently, the maximum of $\kappa_N(\eta)$ is κ_1 and the minimum is κ_2. These are called the *principal curvatures* of the surface $\mathcal{S}$ at P. The directions corresponding to the angles η_1, η_2 are called the *principal directions* at P. As we have just seen, they are $\pi/2$ radians apart.

We may summarize the foregoing results as follows: If the point P is not an umbilic, there exist two normal sections, ninety degrees apart, such that the curvature of one of these sections is greater than the curvature of any other normal section through P, while the curvature of the other one is less than the curvature of any other normal section. At an umbilic, the curvature of every normal section has the same value–both the maximum and minimum values of $\kappa_N(\eta)$ (*i.e.*, the principal curvatures) are then equal to this value. Every direction at an umbilic is a principal direction.

A surface curve whose tangent at each point lies along a principal direction is called a *line of curvature*. The lines of curvature may be exploited to construct a convenient orthogonal system of Gaussian coordinates on $\mathcal{S}$.

Experiment 51 (Principal curvatures): **(a)** For the cylinder that you used in Experiment 48, identify the principal directions and evaluate the principal curvatures. Draw lines of curvature on the cylinder.

(b) Estimate the principal curvatures of a point on an eggplant, or large potato, or watermelon.

(c) Draw a closed curve around the waist of a banana and mark four points, about 90° apart, on the curve. Note that one principal curvature remains approximately constant as you traverse the closed curve. Study how the sign of the other principal curvature changes around the circuit. Can you identify the direction for which this principal curvature is zero? Draw lines of curvature on the banana. □

In Chapter 13, we saw that some geometrical properties of a surface are intrinsic, and hence remain unaltered during bending of the surface, whereas other properties are extrinsic. What about the principal curvatures?

Experiment 52 (Principal curvatures are extrinsic): **(a)** Take a sheet of paper and bend it into a cylindrical or conical shape. Describe what happens to the principal curvatures at a point of the sheet process. Argue that principal curvatures are not intrinsic quantities.

(b) For the cut ball that you had in Experiment 44, examine how the principal curvatures at a point change as you bend the ball. Again, argue that principal curvatures are extrinsic. □

Suppose that we choose lines of curvature as our Gaussian coordinate system. At every point that is not an umbilic, these lines are perpendicular to each other, and at an umbilic, we can pick any two directions that are perpendicular to each other. Hence, by virtue of Equation (12.54), $a_{12} = 0$ for the lines of curvature system. The principal curvatures can be found from Equations (15.13):

$$\kappa_1 = \frac{b_{11}}{a_{11}}\,, \quad \kappa_2 = \frac{b_{22}}{a_{22}}\,. \tag{15.26}$$

If we take unit vectors $\mathbf{e}_1$ and $\mathbf{e}_2$ along the lines of curvature at P, the tangent vector $\mathbf{t}$ is given as in Equation (15.14), with η now measuring the angle that RS makes with the first direction of principal curvature. When $\mathbf{e}_1, \mathbf{e}_2$ are employed as basis vectors, Equations (15.26) reduce to

$$\kappa_1 = b_{11}\ ,\ \kappa_2 = b_{22}\ , \tag{15.27}$$

since a_{11} and a_{22} are the squares of the magnitudes of the basis vectors (see Equations (12.50)) and these are now unity. Substituting Equation (15.27) in Equation (15.15), we may express the curvature of an arbitrary normal section through P as

$$\kappa_N(\eta) = \kappa_1 \cos^2 \eta + \kappa_2 \sin^2 \eta + b_{12} \sin 2\eta\ . \tag{15.28}$$

If P is an umbilic, then κ_N is constant ($= \ \kappa_1 = \kappa_2$), and therefore Equation (15.28) implies that $b_{12} = 0$. If P is not an umbilic, κ_1 cannot be equal to κ_2. Equations (15.27) and (15.21) then imply that at the maximum or minimum values of $\kappa_N(\eta)$,

$$\tan 2\eta \ = \ \frac{b_{12}}{\frac{1}{2}(\kappa_1 - \kappa_2)}\ . \tag{15.29}$$

But, we know that $\kappa_N(\eta)$ takes on its maximum value κ_1 at $\eta = 0$. Consequently, for our choice of $\mathbf{e}_1$ and $\mathbf{e}_2$ as basis vectors, $b_{12} = 0$. Therefore, Equation (15.28) can always be written in the form

$$\kappa_N(\eta) = \kappa_1 \cos^2\eta + \kappa_2 \sin^2\eta\ . \tag{15.30}$$

This beautiful result is known as *Euler's theorem.* It shows that the curvature of normal sections at a point varies in a surprisingly simple way as the sectioning plane is rotated about the surface normal.

Experiment 53 (Euler's theorem): Cut several normal sections from a cardboard cylindrical container and verify that their curvatures are given by the formula $\kappa_N = \ (\cos^2 \eta)/R$. □

The average value of the principal curvatures at a point on a surface is called the *mean curvature*, and is denoted by the symbol H:

$$H = \frac{1}{2}(\kappa_1 + \kappa_2) \,. \tag{15.31}$$

A sphere of radius R has mean curvature $1/R$ everywhere, a cylinder of radius R has mean curvature $1/(2R)$ everywhere, and a plane has zero mean curvature everywhere. Since a piece of the plane can be bent to form a piece of cylinder, it follows that H is not an intrinsic quantity.

For physical reasons, soap bubbles have equilibrium shapes of constant nonzero mean curvature, and soap films tend to have shapes of zero mean curvature. The vanishing of H is closely related to a classical minimization problem called *Plateau's Problem*,[2] which may be stated as follows: on a given closed curve, can a surface be formed whose area is less than or equal to that of every nearby surface, and if so, what is its shape? (By nearby surfaces, we mean surfaces that can be obtained by small deformations of the surface, keeping the bounding curve fixed.) It can be proved that if such a surface does exist, its mean curvature is necessarily zero. Problems related to surfaces of constant (including zero) mean curvature are topics of current mathematical research.[3]

Experiment 54 (Soap films II): **(a)** Make a wire frame consisting of two circles a few inches in diameter, and a handle joining them together (Fig. 117). Dip the circles in a soap solution (see Experiment 33) and withdraw them. Study the shapes of the various surfaces that you get. Move the circles apart slightly and see what happens. Puncture all of the horizontal surfaces that may have formed and study the remaining surface. Does its area appear to be smaller than that of other nearby surfaces?

(b) Make a wire frame having the shape shown in Fig. 118; bend the frame about the handle EF so that AB and DC form almost complete circles in horizontal planes. Study the several different types of soap films

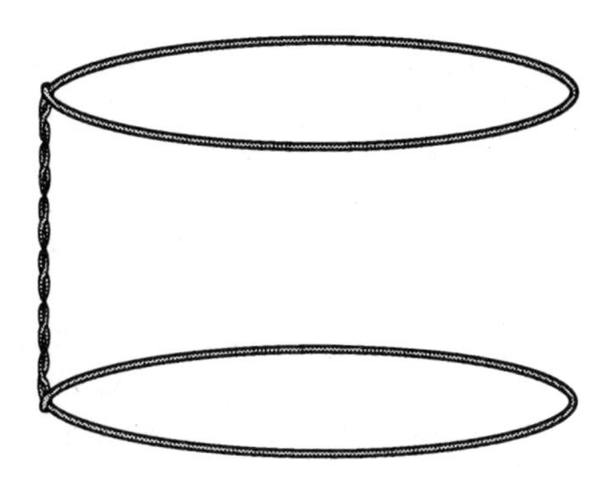

Figure 117 A wire frame

that can be formed on this frame. (Keep your apparatus and solution for use in Experiment 60). □

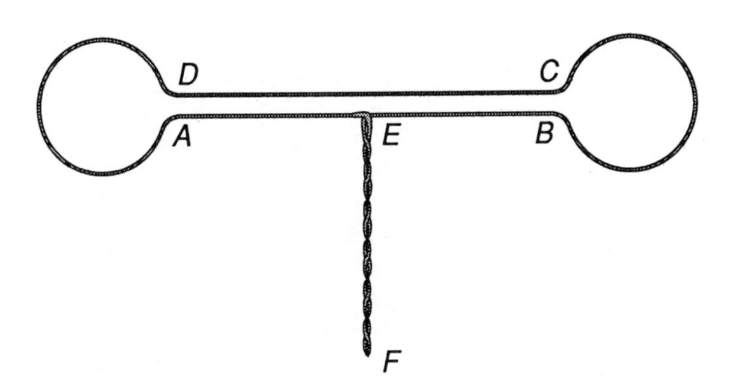

Figure 118 Another wire frame

Notes to Chapter 15

[1]If we are given a nonorientable surface, such as a Möbius strip (Fig. 86), we can always take an orientable piece of it.

[2]J.A.F. Plateau (1801-1883) was a Belgian physicist. Although orphaned at the age of fourteen, he went on to become a professor at the University of Ghent. As a result of an optics experiment in 1829, during which he stared at the sun for 25 seconds, his eyesight was damaged, and he eventually became blind in 1843. Despite this, he continued to experiment on the forms realized by weightless masses of liquid and soap films. (See Elaine Koppelman's entry on Plateau in the *Dictionary of Scientific Biography*).

[3]See again Courant (1940), and Courant and Robbins (1941; pp. 385-397). Advanced treatments are given by Almgren (1966), Osserman (1986), and by Fomenko and Tuzhilin (1991).

16

Gaussian Curvature

In Chapter 15, we saw how, at a point on a surface, the curvature of a normal section varies as the sectioning plane is rotated about the normal vector. We learned that this variation is governed by Euler's formula (15.30). In the present chapter, a completely different approach is taken, which is not based at all on the curvature of curves. Here, we study a brilliant idea of Gauss's, which will enable us to define a unique value of *surface curvature* at each point on a smooth surface.

Gauss's Definition of Surface Curvature

Gauss's celebrated paper "General Investigations of Curved Surfaces," presented in Latin to the Royal Society of Göttingen on 8 October 1827, grew out of his extensive practical work on geodesy.[1] He gradually became concerned with the general mathematical problem of mapping one curved surface upon another in such a way that "the smallest elements remain unchanged". At the heart of the subject is Gauss's innovative definition of surface curvature, which will now be described.

Consider a smooth orientable surface S that, additionally, is both simply connected and arcwise connected (Fig. 119a). Pick one of the two possible unit normal vector fields, and call it the "outward" normal field.

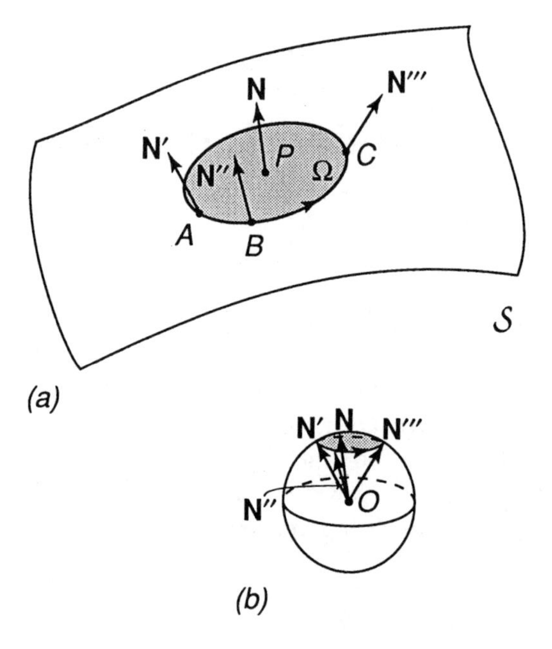

Figure 119 The Gauss map

For any given point P of $\mathcal{S}$, draw a simple closed curve enclosing a region about P: call this region Ω. Let $\mathbf{N}$ be the outward unit normal at P. Traverse the boundary of Ω in the counterclockwise (or positive) direction, as indicated in Fig. 119a. Let $\mathbf{N}'$, $\mathbf{N}''$, $\mathbf{N}'''$ be the surface normals at typical points A,B,C on the boundary. Gauss observed that if an auxiliary unit sphere is introduced, each of the normals $\mathbf{N}$, $\mathbf{N}'$, $\mathbf{N}''$, $\mathbf{N}'''$ (*etc.,*) will uniquely determine a radius vector of the sphere (Fig. 119b). Moreover, as the boundary of Ω is traversed, a closed curve will be cut out on the auxiliary sphere by the tips of the normal vectors emanating from O. This method of assigning points on the unit sphere to points on a surface is called the *Gauss map*.

Let a be the area of the region Ω and let $\bar{a}$ be the area of the cap cut out on the sphere as we go around the boundary of Ω. Reckon $\bar{a}$ as positive if the curve on the sphere is traversed in the positive direction, and negative in the opposite case. If the direction of the normals remains constant as we traverse the boundary of Ω, then $\bar{a}$ will vanish: this happens in the case of a plane. Roughly speaking, the less Ω differs from a piece of the plane, the smaller $\bar{a}$ will be. With characteristic insight, Gauss saw that $\bar{a}$ could be regarded as a measure of the curvature of the region: he called it the *total curvature* of Ω. Thus,

$$\textit{total curvature of } \Omega = \bar{a} \ . \tag{16.1}$$

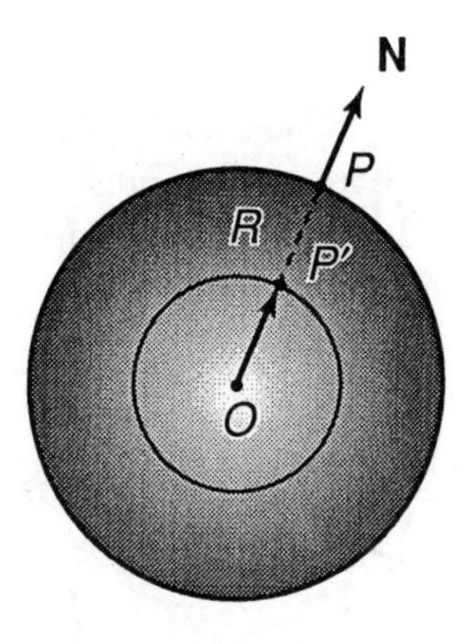

Figure 120 The Gauss map for a sphere of radius R

Example 1: A special spherical segment. Consider the spherical segment bounded by latitudes $\phi = 0$ and ϕ, and longitudes $\theta = 0$ and θ, which we met previously in Example 5 of Chapter 12 (p.186). The Gauss map is easy to construct: draw a unit sphere that is concentric with the given sphere of radius R; for any point P on the latter, draw a radius vector OP and let the reference plane intersect the unit sphere in

P' (Fig. 120); P' is the image of P under the Gauss map.

Equation (12.60) tells us that the area of the spherical segment is $R^2\theta \sin \phi$. The corresponding area cut out on the unit sphere is $\theta \sin \phi$. Therefore, the total curvature of the spherical segment is $\theta \sin \phi$. In particular, the total curvature of a hemisphere is $2\pi \sin (\pi/2) = 2\pi$.

Experiment 55 (Total curvature): (a) On a cardboard cylindrical container, enclose a small region by a closed curve as in Fig. 119(a). Track the outward surface normals by thumbtacks or toothpicks. Use a ball (or round balloon) to represent the auxiliary sphere. Identify the points on the ball that the Gauss map assigns to the points on the cylinder. Evaluate the total curvature of the cylinder.

(b) Repeat this process for a small region marked out on a watermelon or eggplant.

(c) Consider the two surfaces in Fig. 121: one has the shape of a pressure vessel with hemispherical ends; the other is egg-shaped. Use a round balloon to represent the auxiliary sphere, and ink in the area that would be cut out on the balloon by the Gauss map of the surface normals along the curves *ABCD*. Evaluate the total curvature of the entire surface

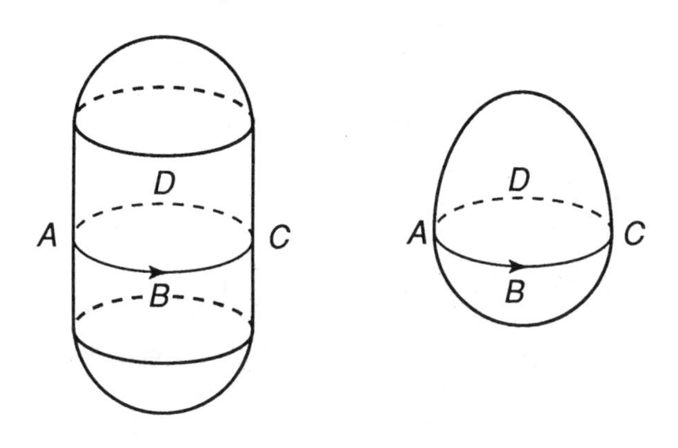

Figure 121 A pressure vessel and an egg

of the pressure vessel and of the egg. Under which class of deformations of a sphere does the total curvature remain unaltered? What is the total curvature of a box (Fig. 84)? □

It is evident from Experiment 55 that two regions of a surface may be curved in very different ways and yet have the same total curvature. Another example of this is shown in Fig. 122: The hat and the flat piece of material from which it was deformed have the same total curvature (zero). Paralleling what we did for curves in Chapter 10, let us compare

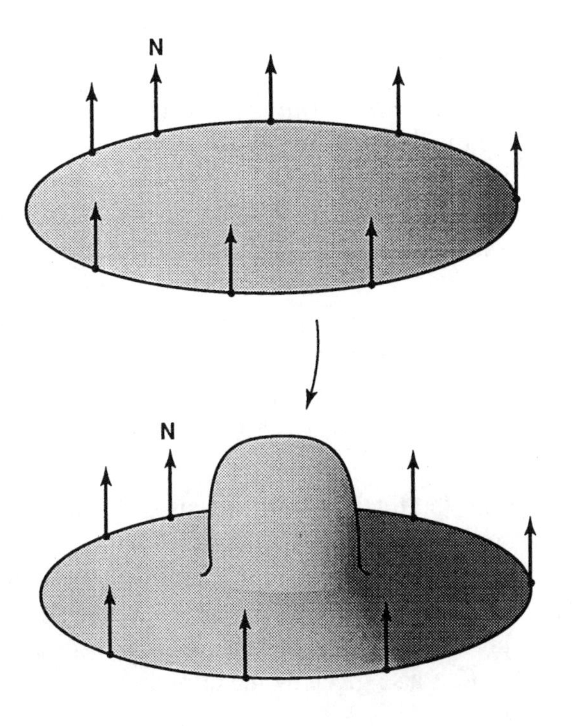

Figure 122 A hat

total curvatures over equal areas. Thus, we define an average curvature for surfaces by

$$average\ curvature\ of\ \Omega = \frac{\bar{a}}{a}. \qquad (16.2)$$

For the spherical segment in Example 1, the average curvature is $\theta \sin\phi$ divided by $R^2\theta \sin\phi$, or $1/R^2$.

Gauss took the dramatic step of defining a density of curvature, which he called the *measure of curvature of a surface at a point P*, by taking the limit of the average curvature of Ω as the area of Ω tends to zero:

$$\lim_{a\to 0} \frac{\bar{a}}{a} = K. \qquad (16.3)$$

Mathematicians now call K the *Gaussian curvature*. For the important special case of the Euclidean plane, $K = 0$ everywhere. For a sphere of radius R, the Gaussian curvature is $1/R^2$: smaller spheres have larger Gaussian curvatures. On a saddle, or mountain pass, K is negative. For each of the two surfaces in Fig. 121, K is positive everywhere (but it is not constant).

When the limit (16.3) exists everywhere in Ω, we may express the total curvature of Ω as the surface integral

$$\bar{a} = \int_\Omega K\, da, \qquad (16.4)$$

and for this reason, the total curvature is also known as the *integral curvature* of the region.

It is difficult to measure the areas a and $\bar{a}$ accurately, and consequently the formula (16.2) does not furnish an easy means of estimating the Gaussian curvature at a point on a general physical surface. However, it can be shown that K is also given by the formula[2]

$$K = \frac{b_{11}b_{22} - b_{12}^2}{a_{11}a_{22} - a_{12}^2}, \qquad (16.5)$$

where a_{11}, a_{12}, a_{22} are the coefficients of the first fundamental form [see Equation (12.27)] and b_{11}, b_{12}, b_{22} are the coefficients of the second fundamental form [see Equation (15.11)]. If unit vectors $\mathbf{e}_1, \mathbf{e}_2$ are chosen to lie along the principal directions at P, then as we saw near the end of Chapter 15, $a_{11} = a_{22} = 1, a_{12} = 0$, and $b_{12} = 0, b_{11} = \kappa_1, b_{22} = \kappa_2$. It then follows immediately from Equation (16.5) that

$$K = \kappa_1 \kappa_2 \,, \tag{16.6}$$

i.e., the Gaussian curvature is equal to the product of the principal curvatures. This is an extremely important, and very useful result. It is striking that surface curvature can be expressed as the product of the curvatures of two curves. On a cylinder, K must vanish, since one of the principal curvatures is always zero, while on a sphere, $K = (\frac{1}{R})(\frac{1}{R}) = \frac{1}{R^2}$.

Consider a saddle (Fig. 123). At a point such as P, the principal curvature κ_1 is positive, being that of the normal section AB, whereas the principal curvature κ_2 is negative (being equal to that of CD). The Gaussian curvature at P is therefore negative. On a bicycle tube (or

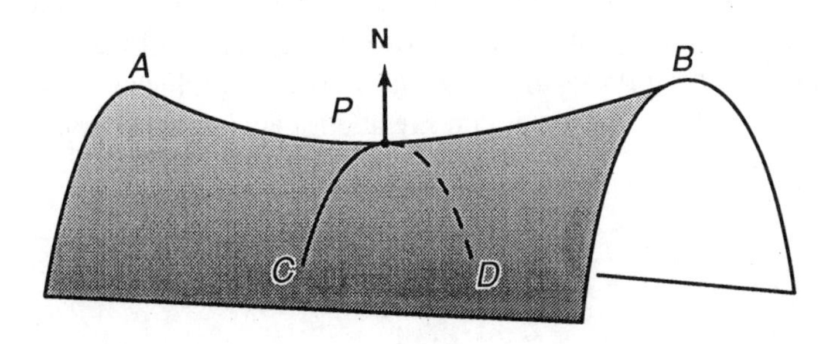

Figure 123 A saddle

torus), K is positive on the very outside, negative on the very inside, and vanishes on two circles lying in between.

Experiment 56 (Evaluating Gaussian curvature): **(a)** Evaluate the Gaussian curvatures of a basketball, a tennis ball, and a golf ball.

(b) Take an eggplant (or two identical ones), or any other suitable vegetable or fruit. Choose any convenient point on them. Identify the principal directions at the point. Measure the principal curvatures (see Experiment 23). Calculate the Gaussian curvature. □

Many common surfaces have points (either vertices or points belonging to edges) where a unique normal vector does not exist. Strictly speaking, we cannot apply the Gauss map at such points, nor does there exist a value of K at them. For physical objects, we can always imagine the edges and vertices as being rounded off, and we may thereby evaluate the total curvature of a region enclosing a vertex or an edge. Suppose for example that the tops of the objects in Fig. 121 become progressively more pointed, while the remainder of the objects, including the curve $ABCD$, are left unchanged. The total curvature of the top portions, evaluated from normals constructed around $ABCD$ will remain the same as before, but as the tops become vertices, K will tend to infinity there. Likewise, as the hat in Fig. 122 is deformed into a witch's hat, K tends to infinity at the apex.

Experiment 57 (Total curvature of pieces of a box): Again using a round balloon to represent the auxiliary sphere, evaluate the total curvature of the surfaces *EFGHIJ* and *KLMNOPQ* which you prepared in Experiment 28. What can you say about the Gaussian curvature for these surfaces? □

The Remarkable Theorem

In Chapter 13, it was pointed out that surfaces have an inner or intrinsic geometry that is independent of their shape in space, and can be determined by observations made on the surface itself. We have seen in Experiment 52 that the principal curvatures, κ_1 and κ_2, are not intrinsic quantities: they can be changed by bending a surface. Looking at Equation (16.6), one would suspect that the Gaussian curvature K might also be extrinsic. However, in his great memoir of 1827, Gauss proved that K can be calculated from the coefficients a_{11}, a_{12}, a_{22} of the first fundamental form [see Equation (12.27)] and their first and second partial derivatives. In other words, he discovered that K is an intrinsic quantity! Its value can be determined from surface measurements alone. Since surface measurements are not altered by bending, Gauss had now arrived at what he himself considered to be a "remarkable theorem" (theorema egregium):[3]

> If a smooth curved surface, or any part of it, can be developed upon another surface, the Gaussian curvature at every point remains unaltered.

If follows as a corollary of the Theorema Egregium that if a smooth surface can be developed upon the plane, then $K = 0$ throughout the piece. On the other hand, if K is different from zero at some point of the piece, then the piece is not developable. For example, no segment of a sphere is developable (since $K = 1/R^2 > 0$).

It can be shown that if $K = 0$ everywhere on a smooth orientable surface, then the surface is locally developable at each of its points. In other words, at each point P of the surface, some finite piece of the surface in the vicinity of P can be bent so that it coincides with a region of the Euclidean plane. The local intrinsic geometry of the surface is identical to the local geometry of the Euclidean plane. If K is not zero everywhere on a piece of surface, then the local surface geometry cannot be Euclidean. The Gaussian curvature is essentially a measure of the local non-Euclideanness, or non-flatness, of a surface.

Experiment 58 (Gaussian curvature and developability): **(a)** Discuss the Gaussian curvature for points on a cylindrical surface, and for points on the cowl in Experiment 43.

(b) Explain why the surface $KLMNOPQ$ in Experiment 28 cannot be developable.

(c) For the cut ball that you used in Experiments 44 and 52, try to convince yourself that as you bend the ball, the principal curvatures vary in such a way that their product is constant. □

For surfaces whose mean curvature H is zero, it follows from Equations (15.31) and (16.6) that

$$K = -\kappa_1^2 \,. \tag{16.7}$$

Consequently, such surfaces cannot have positive Gaussian curvature.

Experiment 59 (Soap films III): Examine how the shape of a soap film changes from point to point in a manner that keeps its Gaussian curvature negative or zero. □

Notes to Chapter 16

[1] A translation is given in Gauss (1827).

[2] See p. 83 of Struik (1961). Struik uses the symbols E, F, G and e, f, g to denote the coefficients of the first and second fundamental forms, respectively.

[3] The theorem can be found on p. 20 of Gauss (1827). See also Gauss's Abstract on pp. 45-49 of Gauss (1827).

17

Riemann (1826-1866)

Like Mozart's, Bernhard Riemann's life was short but marvelously creative. He solved several of the most difficult problems in pure and applied mathematics, introduced entirely new concepts and techniques, and profoundly changed the way in which mathematicians, physicists, and philosophers view space.[1]

The Riemann family lived in a village in Hanover, Germany. Georg Friedrich Bernhard, the second of six children, was born on 17 September 1826. The family was tightly knit and very supportive, but hardly affluent. Early education of the children was provided by their father, who was a Lutheran minister. Bernhard's mathematical ability soon asserted itself: he was gifted at arithmetic and showed intense interest in mathematics at school.

In 1846, Riemann became a student of theology and philology at Göttingen University, but he also attended lectures in mathematics and physics, including a course from Gauss. He soon realized that he wanted to be a mathematician, and obtained his father's approval to change his course of studies. Also, in conjunction with this decision, Riemann moved to the University of Berlin in 1847, where a group of brilliant young mathematicians were teaching. He learned much during his two years' stay, and was especially influenced by Dirichlet (1805-1859). At this stage, Riemann was already developing original ideas on the theory of complex variables.

Riemann returned to Göttingen in 1849 to work towards his doctoral degree. The physicist, Wilhelm Weber (whom we already met in Chapter 14), was now also back in Göttingen, and Riemann benefited much from his presence. For his doctoral dissertation, he chose a fundamental topic, the foundations of the general theory of functions of a complex variable.

He presented his work to Gauss in 1851, who immediately recognized its supreme originality. Riemann's approach to complex analysis had a distinctly conceptual and geometric flavor rather than the prevailing calculational or algorithmic one.

Riemann had a university career in mind, and having obtained his doctoral degree, proceeded towards the next step, which was that of *Privatdozent*, or unsalaried lecturer. A Privatdozent was permitted to conduct university courses and to earn fees from the students who attended. To become a *Privatdozent,* one had to prepare another dissertation (*Habilitationsschrift*) and, in addition, one had to present a lecture (*Habilitationsvortrag*) before the faculty. Between 1851 and 1853, Riemann worked on a variety of topics. He submitted a memoir on the representability of functions by means of trigonometric series. It was here that he introduced the definition of an integral which has been named after him and is taught in every calculus course. The subject chosen by Riemann for his *Habilitationsschrift* is an unusually rich one, and later led to major new developments in our understanding of the infinite, and in further elaboration of the concept of an integral.

Next, Riemann submitted the titles of three possible topics for his *Habilitationsvortrag*. From these, the faculty would choose the one on which he should present his lecture. The first two were on mathematical questions related to physical problems; these he had worked out in detail. The third, on the foundations of geometry, he had not yet prepared. To Riemann's surprise, the faculty, at Gauss's recommendation, chose the third topic. Gauss himself had long pondered upon the foundation of geometry, and he must have had sufficient belief in Riemann's powerful imagination to trust that the young man would create sublime mathematics from this extremely difficult, but enormously rich material. Although Riemann regarded Gauss's choice as a trial, it really was a gift from the aged Gauss to him, a gift which he was to repay in excess of all expectations.

Busy with other work, Riemann only began preparing his lecture in the spring of 1854. He spent about seven weeks on it, and arranged to deliver it on Saturday, 10 June 1854, starting at 11:30 a.m. His audience consisted of classicists and philosophers as well as mathematicians and

Bernhard Riemann (1826 - 1866)

physicists. Gauss, already ailing from heart disease, listened attentively. Riemann's lecture, entitled "On the Hypotheses which lie at the Foundations of Geometry", starts out thus:[2]

> *It is known that geometry assumes, as things given, both the notion of space and the first principles of construction in space. She gives definitions of them which are merely nominal, while the true determinations appear in the form of axioms. The relation of these assumptions remains consequently in darkness; we neither perceive whether and how far their connection is necessary, nor, a priori, whether it is possible.*
>
> *From Euclid to Legendre (to name the most famous of modern reforming geometers) this darkness was cleared up neither by mathematicians nor by such philosophers as concerned themselves with it.*

Riemann explains that the source of the difficulty is that it had not been previously realized that one can construct a concept of space which is completely independent of any notion of distance. Nowadays, students of mathematics are accustomed to deal with sets of the most abstract character, and to define unconventional measures of distance between the elements of them, but this progress has come about only because of the monumental efforts of such mathematicians as Riemann and Cantor.

Riemann broke with the prevailing philosophical and mathematical traditions by suggesting that in the concept of space no definite measure of distance is implied. In modern terms: the only definite notions embodied in the concept of space are topological ones. As a consequence, different metrics can be imagined, and in order to determine whether space is Euclidean or not, one must make physical observations. The implications of Riemann's viewpoint for geometry, physics, and the interconnection between the two fields, are exceedingly profound.

It was in Riemann's lecture that several vital concepts of modern mathematics and physics appeared for the first time. In particular:

(a) he introduced the concept of a manifold;

(b) he explained how different metric relations could be defined on a manifold;

(c) he extended Gauss's notion of intrinsic curvature of a surface to higher dimensional manifolds;

(d) he drew the distinction between the properties of infiniteness and unboundedness of a space; and

(e) he asserted that the geometry of space can only be decided by experiment, and even suggested that the metric relations of space might be connected with the forces that somehow hold space together.

Let us elaborate on these ideas.[3] As for item (a), we note first that curves and surfaces are examples of one-dimensional and two-dimensional manifolds, respectively. Recall that the essential feature of a curve is that, for every piece of it, points can be located by a single parameter, or coordinate. Likewise, for each piece of a surface, points can be located by means of two Gaussian coordinates each of which runs through an interval of real numbers. Generalizing, we say that an n-dimensional manifold is a set, such that for every piece of it, we can locate points by using n coordinates each of which runs through an interval of real numbers.[4] Thus, as Riemann says, the essential character of a manifold is that its points can be located through numerical specifications.

Just as with curves and surfaces, certain geometrical properties of manifolds can be investigated even though no notion of distance is present. For instance, questions such as the connectivity and the orientability of a manifold can be dealt with. Multidimensional manifolds appear constantly in modern mathematics, physics, and engineering. For example, the possible configurations of a spinning rigid body, the thermodynamic states of gas, and the states of a dynamical system, can each be represented by points in a manifold of appropriate dimension.

Regarding item (b), Riemann takes a clue from Gauss's celebrated *Disquisitiones Generales Circa Superficies Curvas*, which was discussed in Chapter 16. Consider a point P in an n-dimensional manifold and let $u_1, u_2, ..., u_n$ be its coordinates (Fig. 124). Take a second point Q whose coordinates $u_1 + du_1, u_2 + du_2, ..., u_n + du_n$ differ only infinitesimally from those of P. Riemann suggests as one possible definition that the square of the length ds of the line element joining P to Q is given by

$$ds^2 = \sum_{i=1}^{n} \sum_{j=1}^{n} g_{ij} \, du_i \, du_j \ , \tag{17.1}$$

where the numerical quantities $g_{ij}(i = 1, 2, ..., n; \ j = 1, 2, ..., n)$ are functions of $u_1, u_2, ..., u_n$. This directly generalizes to an n-dimensional manifold the formula (12.27) which Gauss obtained for the line element of a surface (*i.e.*, a two-dimensional manifold). The expression on the right-hand side of equation (17.1) is a quadratic form in the variables $du_1, du_2, ..., du_n$. According to our usual notion of length, ds^2 is positive unless Q and P coincide. Accordingly, the quadratic form is said to be *positive definite*. The rule (17.1) for determining length is called a *Riemannian metric*. A manifold furnished with a Riemannian metric is called a *Riemannian manifold* or a *Riemannian space*. The intrinsic geometry of a Riemannian space is defined by the metric (17.1). Since the metric can change from point to point in the space, it is evident that very different intrinsic geometries can exist in different regions of the space.

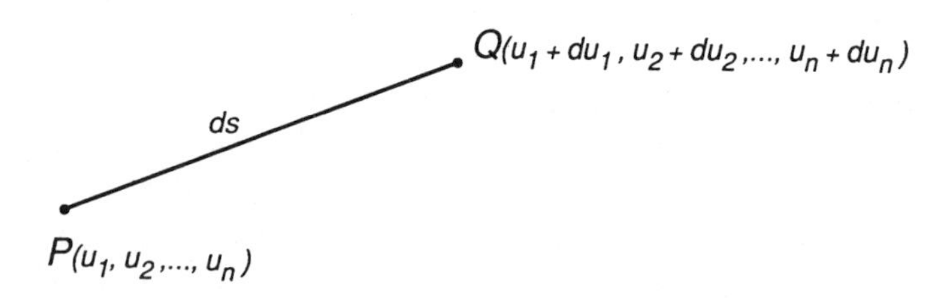

Figure 124 An infinitesimal line element joining two points in a manifold

In an n-dimensional Euclidean space, the square of the length of a line joining any two points is given by the Pythagorean formula. If the points are infinitesimally close to one another, we obtain

$$ds^2 = dx_1^2 + dx_2^2 + \dots + dx_n^2 = \sum_{i=1}^{n} dx_i^2 \,, \qquad (17.2)$$

where $x_1, x_2, \dots, x_n$ are rectangular Cartesian coordinates. It is clear that (17.2) is a special case of (17.1) with

$$\begin{aligned} du_1 &= dx_1,\ du_2 = dx_2,\ \dots,\ du_n = dx_n \,, \\ g_{11} &= 1,\ g_{22} = 1,\ \dots,\ g_{nn} = 1 \,, \qquad (17.3) \\ g_{12} &= g_{21} = g_{13} = g_{31} = \dots = g_{in} = g_{ni} = 0 \,. \end{aligned}$$

Thus, Euclidean space is a very special kind of Riemannian space.

Now, it can be shown mathematically that, at any point P in a Riemannian space, it is always possible to find a set of coordinates (say $\overline{u}_1, \overline{u}_2, \dots, \overline{u}_n$) such that the expression (17.1) reduces to

$$ds^2 = \sum_{i=1}^{n} d\overline{u}_i^2 \,, \qquad (17.4)$$

which has the same form as equation (17.2). If we move on to another point in the manifold, an equation of the type (17.4) will again hold, but with respect to a set of coordinates which are in general different from $\overline{u}_1, \overline{u}_2, \dots, \overline{u}_n$. Thus, in a Riemannian space, it is necessary in general to change coordinates continually if one wishes to obtain the reduction (17.4). It is only in a Euclidean space that a set of coordinates exists such that with respect to this same set, the quantity ds^2 is given by equation (17.2) at *all* points of the space. Riemann calls Euclidean spaces *flat*. The fact that it is possible to reduce (17.1) to (17.4) at any given point of a Riemannian space implies that as far as calculating length is concerned, in the infinitesimal neighborhood of each of its points, a Riemannian space appears to be Euclidean. Just as we may think of a plane curve as being approximated by its family of tangent lines, we may think of a Riemannian space as being approximated by a collection of Euclidean spaces. We may say that a Riemannian space is infinitesimally flat.

To find the length $l\ (C)$ of a curve C joining two arbitrary points in a Riemannian space, we integrate the expression for ds along the curve:

$$l(C) = \int_C \sqrt{g_{ij}\, du_i\, du_j}\ . \tag{17.5}$$

You may think of this process as adding up the lengths of infinitely many line elements into which the curve has been partitioned. The distance between two points can now be defined as the greatest lower bound of the lengths of the curves which join the two points, or roughly speaking, as the length of the shortest curve joining the two points.

Item (c), on curvature, has to do with the question of how one might characterize the deviation from flatness that can occur in a Riemannian space. Here again, Riemann found the key to the solution of this fundamental problem in Gauss's *Disquisitiones Generales Circa Superficies Curvas*. As we saw in Chapter 13, Gauss was the first person to recognize the distinction between properties of a surface which depend upon the shape in which it is given in space, and more deeply-rooted properties of it which are independent of the various shapes into which a surface can be bent without stretching (or tearing, or overlapping). This is the distinction between what we have referred to as extrinsic and intrinsic properties of the surface. Gauss showed that the Gaussian curvature K is an intrinsic measure. Likewise, Riemann sought a quantity by which to characterize the intrinsic geometry of a Riemannian space. The resulting measure is defined by a complicated mathematical object which has several components at each point of a multidimensional manifold. In a Euclidean (or flat) space, the Riemann curvature is zero everywhere. Conceptually, the Riemann curvature is a measure of the degree to which a Riemannian space differs from a Euclidean space. The curvature can vary from point to point, but there are important special cases in which Riemann's measure is constant across the entire space.

Proceeding to item (d), it may appear strange at first that a space could be finite but unbounded. Riemann believed that three-dimensional space is unbounded (*i.e.,* that it has no boundary). But, he also recognized that this does not imply that it is infinite in size. The sphere, for example, is a two-dimensional manifold without a boundary, and yet its area is finite. Riemann saw the possibility of having a three-dimensional space which is curved in such a way that it is finite but has no boundary.

Turning now to item (e), recall that Riemann, through his general concept of a manifold, had succeeded in peeling away the metric overlay to reveal the essential topological nature of space. An abstract three-dimensional space can be endowed with infinitely many different choices of metric, and in order to find out which one is obeyed by the space we live in, we must make physical observations.

Towards the end of his lecture, Riemann makes the astonishing suggestion that the actual metric, and hence the intrinsic geometry, of our space is related to the forces that keep space intact. The full impact of this premonition would only become manifest more than half a century later, when Einstein would make it the philosophical cornerstone of his General Theory of Relativity.

Einstein, above all others, appreciated Riemann's profound insight into geometry. He often expressed his debt to Riemann, and on one occasion, he wrote:[5]

> *Only the genius of Riemann, solitary and uncomprehended, had already won its way by the middle of the last century to a new conception of space, in which space was deprived of its rigidity, and the possibility of its partaking in physical events was recognized.*

It is unlikely that many in the audience on that historic June morning in 1854 appreciated the depth of Riemann's lecture. Evidently, a few decades were to pass before the revolutionary character of the ideas which he propounded was generally recognized.[6] However, the most important member of the audience, Gauss, was astonished by what he had just heard: on the way back from the lecture with his friend Wilhelm

Weber, he spoke, with uncharacteristic enthusiasm, about the profundity of Riemann's ideas.

Riemann was granted the right to lecture, and was overjoyed at the unexpectedly large audience of eight who attended his first class. The great Gauss died in February 1855, and his position was given to Dirichlet, who was a good friend of Riemann. Riemann continued to work strenuously on difficult questions in both pure and applied mathematics. He was advanced to the level of assistant professor in 1857.

Dirichlet died in May 1859, and Riemann was chosen to succeed him as full professor. At this stage of his career, his reputation as a mathematician had spread throughout Europe, and he was on friendly terms with many gifted mathematicians in Germany, Italy, and France. He visited Paris in 1860 and was warmly received by his French colleagues.

In June 1862, Riemann married Elise Koch, a friend of his family. But, in July 1862, he fell ill with pleurisy. He never fully regained his health after this attack. (It has been suggested that he suffered from tuberculosis.) Seeking a cure in a better climate, Riemann spent the winter of 1862-1863 on the island of Sicily. In the spring of 1863, he toured Italy and took great pleasure in the artistic heritage of the several cities which he visited. He returned to Göttingen, but again his health worsened. Another trip to Italy ensued, being brightened this time by the birth of a daughter, Ida Riemann, in Pisa. The winter of 1865-1866 was spent back in Göttingen. Riemann continued to work whenever he could. In June 1866, he went to northern Italy once more with the hope of recovering his strength. Unfortunately, his health deteriorated rapidly and the brilliant Riemann died peacefully and, in full consciousness, on 20 July at Selasca on Lake Maggiore.

Notes to Chapter 17

[1] An insightful account of Riemann's life and work is given by H. Freudenthal in the *Dictionary of Scientific Biography*. At the end of Riemann's collected works (Riemann, 1876), a biographical essay by Dedekind can be found. In a recent book, Monastyrsky (1987) provides details of Riemann's life as well

as an extensive description of his mathematical contributions. Le Corbeiller (1954) presents a succinct account of Riemann's geometrical ideas and their roots. Bell (1937) and Simmons (1992) have interesting chapters on Riemann. See also Chapter 12 of Gray (1979).

[2] Riemann's lecture (which contains only a few equations) is reprinted in his mathematical papers (Riemann, 1876). Translations of it are given in:
(1) Clifford (1882), pp. 56-71; (2) Smith (1929), Vol. 2, pp. 411 - 425;
(3) Spivak (1970) , Vol. 2, pp. 4A- 4 to 4A-20; and (4) McCleary (1994), pp. 269-278 [based on Spivak's translation]. Spivak also presents an analysis of the lecture in Chapter 4B of the cited work. All of the translations (not to mention the original German) are difficult to read, but are well worth the effort. The quotation is taken from Clifford's translation.

[3] I am taking the liberty of using modern mathematical terminology in place on some of Riemann's rather heavy philosophical terms. The correspondence with the terminology in the translations cited in Note 2 can be readily supplied by the industrious reader. One should keep in mind that Riemann had to express new concepts in the language which was available to him. It is only as a result of his efforts, and those of later mathematicians, that we now have a simpler language for expressing Riemann's powerful ideas.

[4] Riemann also discusses *discrete* manifolds, the elements of which can be identified by n variables running through a discrete set of numbers (such as the integers).

[5] See Einstein (1954), p. 274. The quotation is from an essay published in 1934.

[6] For a discussion of the reception of new ideas in geometry in the late 19th century, the reader is referred to Freudenthal (1962).

Tullio Levi-Civita (1873-1941)

18

Levi-Civita (1873-1941)

We have seen how Gauss's theory of surface geometry shaped Riemann's conception of general curved manifolds. Riemann's ideas in turn were taken up by Christoffel (1829-1901), Beltrami (1835-1900), and others. During the closing decades of the 19th century, a powerful school of mathematics developed at the University of Padua. It was here that Levi-Civita came into contact with modern geometry.

Tullio Levi-Civita was born in Padua on 29 March 1873.[1] His father, Giacomo, who was a lawyer, also served as mayor of Padua, and as a senator of the Kingdom of Italy. In his youth, Giacomo Levi-Civita had been a Garibaldi volunteer and had fought in the campaign of 1866. He devoted his life to the unification and independence of Italy. The Levi-Civita family was wealthy and politically liberal. Tullio received his early education at home from a Catholic priest and went to secondary school at the age of ten. He was an excellent student in all areas, but exhibited special interest in mathematics. At the age of fifteen, he tried to prove Euclid's parallel postulate, and although his argument was elegant, it had a pernicious flaw. Little did he know that one day his name would be associated with a brilliant extension of the notion of parallelism.

Giacomo had desired for Tullio to become a lawyer, but when it became clear that the son wished earnestly to study mathematics, the father encouraged him and took much pride in his later successes. In 1890, Levi-Civita enrolled as a student of mathematics at the University of Padua. He was greatly inspired by the mathematical physicist Ricci (1853-1925)[2] and also by the geometer Veronese (1854-1917). His first mathematical paper was published in 1893.

Levi-Civita received his degree in 1894 and pursued a teaching career. His beginning appointment was at a teachers' college in Pavia. But soon after, in 1896, he returned to Padua to teach rational mechanics at the University. He spent an extended and productive period there, and eventually became recognized as a leading mathematical physicist and a great teacher.

In personality, Levi-Civita was unpretentious, spontaneous, and an excellent conversationalist. Students and colleagues alike were drawn to him, and his wide knowledge of scientific subjects was often sought after. In physical aspect, he was small, and he suffered from near-sightedness. He was an avid mountaineer and bicyclist. He was also fond of traveling abroad. In 1914, he married Libera Trevisani, who was a doctoral student of his at the University of Padua. The couple were very well-liked in the international mathematical community, and were known for their gracious hospitality to visiting scientists and mathematicians. They did not have any children.

The field of research for which Levi-Civita is best known is that of the *absolute differential calculus*. This branch of mathematics comprises an efficient and extremely powerful apparatus for dealing with a class of functions (later called tensors) which transform in definite ways under general changes of coordinates. It was invented by Ricci in 1884, and was elaborated upon by him for many years. The definitive statement of the theory was published by Ricci in collaboration with Levi-Civita in 1900.[3] In this work, the absolute differential calculus came of age as the natural language for the description of Riemann's profound ideas on the geometry of curved spaces.

During the first decade of the 20th century, only a limited amount of effort was devoted to the calculus of Ricci and Levi-Civita. But then, in 1916, the field emerged as a central part of modern mathematical physics when Einstein found in it the most suitable instrument for expressing his revolutionary ideas on general relativity.

In 1917, Levi-Civita made a fundamental advance in the absolute differential calculus when he invented the concept of *parallel displacement* (or *parallel transport*). This notion, which attested his acute geometrical intuition, will be described in Chapter 19. It is for this particular contribution that Levi-Civita is best known.

After the devastation suffered in Italy during World War I, the University of Rome wished to re-establish itself as a prominent center of learning. Levi-Civita, by then regarded as one of Italy's finest mathematicians, was invited to come to Rome. Although reluctant to leave his native Padua, he agreed to take a position at the University of Rome in 1918, first as professor of higher analysis and subsequently as professor of rational mechanics. His research encompassed a wide variety of topics, including differential geometry, analytical mechanics, celestial mechanics, hydrodynamics, electromagnetism, and relativity. He attracted many students, and by his scientific activities did much to increase the international prestige of Rome, and of Italy in general. He and his wife frequently traveled abroad. He received many honors from foreign, as well as Italian, scientific societies. He was elected a foreign member of The Royal Society of London in 1930. In 1933, Levi-Civita presented a paper on mathematical physics at a joint meeting in Chicago of the American Association for the Advancement of Science and the American Mathematical Society, held in conjunction with the Century of Progress Exposition of that year.

In addition to his scientific papers, Levi-Civita wrote several books. His style is always direct, clear and precise, while maintaining a strong intuitive flavor. In this regard, a remark on geometry with which he opens his *Absolute Differential Calculus* is particularly revealing:[4]

> *In analytical geometry it frequently happens that complicated algebraic relationships represent simple geometrical properties. In some of these cases, while the algebraic relationships are not easily expressed in words, the use of geometrical language, on the contrary, makes it possible to express the equivalent geometrical relationships clearly, concisely, and intuitively. Further, geometrical relationships are often easier to discover than are the corresponding analytical properties, so that geometrical terminology offers not only an illuminating means of exposition, but also a powerful instrument of research.*

During most of his life, Levi-Civita was free from external, and especially political, worries. However, by the mid-1920s, the monster of Fascism was beginning to devour civilized life in Italy. After the assassination of the parliamentary leader of the Socialist party by the Fascists, Levi-Civita, along with many others, signed a document criticizing Fascist practices. For several years thereafter, Levi-Civita was protected by his scientific eminence at home and abroad. He continued to produce new work in mathematical physics, and to enjoy teaching. However, in September 1938, the Fascist regime, under the influence of Nazism, decreed that all professors of Jewish origin in Italy be removed from their positions. Levi-Civita is said to have learned of the anti-Semitic decrees at his sister's house when somebody turned on the radio just as the new proclamations were being broadcast. He appeared calm, but the summary dismissal from the institutions he cherished affected him deeply and soon took an irreversible toll on his health.

The Vatican supported Levi-Civita against Mussolini's government. In January 1940, the Pontifical Academy of Science announced that a scientific congress on the age of the earth was being organized and that Levi-Civita and the other leading Italian mathematician of the period, Vito Volterra (1860-1940), who was also of Jewish origin, were to play a prominent role at the meeting.[5]

Although robust in health during almost all of his life, Levi-Civita became seriously ill after 1938. He was diagnosed as suffering from heart disease and was advised by his doctors to avoid long trips. He stayed in Italy even though several foreign universities offered him asylum. On 29 December 1941, he died from a stroke in Rome. At first, the Italian newspapers, except for the Vatican's *Osservatore Romano*, ignored his death, but were later influenced by the Pontifical Academy to announce it.[6] An obituary which does full justice to the great mathematician and wonderful human being that Tullio Levi-Civita was, was published by The Royal Society of London in November 1942.[7]

Notes to Chapter 18

[1] A biography of Levi-Civita by Ugo Amaldi appears in Vol. 1 of Levi-Civita (1954). Other excellent accounts of his life and work may be found in the following publications: (1) *Obituary Notices of Fellows of The Royal Society*, Vol. 4, 1942, pp. 150-165; (2) *Bollettino della Unione Matematica Italiana*, Series 4, Vol. 8, 1973, pp. 373-390. I have also consulted the (much too brief) article on Levi-Civita, as well as those on Ricci-Curbastro, Veronese, and Volterra, in the *Dictionary of Scientific Biography*.

[2] Ricci-Curbastro is usually referred to simply as Ricci.

[3] This paper, entitled *Méthodes de Calcul Diffentiel Absolu et leurs Applications*, was invited by the mathematician Felix Klein (1849-1925) and appeared in Vol. 54 of the *Mathematische Annalen*. It is reprinted in Levi-Civita (1954), Vol. 1, pp. 479-559. A translation of it, together with a commentary, can be found in Hermann (1975).

[4] See Levi-Civita (1926), p. 1.

[5] This is reported in the *New York Times* of 10 January 1940, p. 19. Volterra, who served Italy loyally during World War I, and who established and took charge of the Italian Office of War Inventions in 1917, also suffered terribly at the hands of Mussolini's regime. (For more information on this, see the article on Volterra cited in Note 1.)

[6] A brief obituary of Levi-Civita appeared in the *New York Times* on 2 January 1942.

[7] This is cited in Note 1 above.

19

Parallel Transport of a Vector on a Surface

In this chapter, we describe a particular way of moving a vector along a given curve on a surface. It provides an especially revealing means of exploring the non-Euclideanness of the surface.

First, let us consider the process of moving a vector parallel to itself along a curve in the plane. Thus, choose any two distinct points A' and B' in the Euclidean plane and join them by an arbitrary curve $A'P'B'$

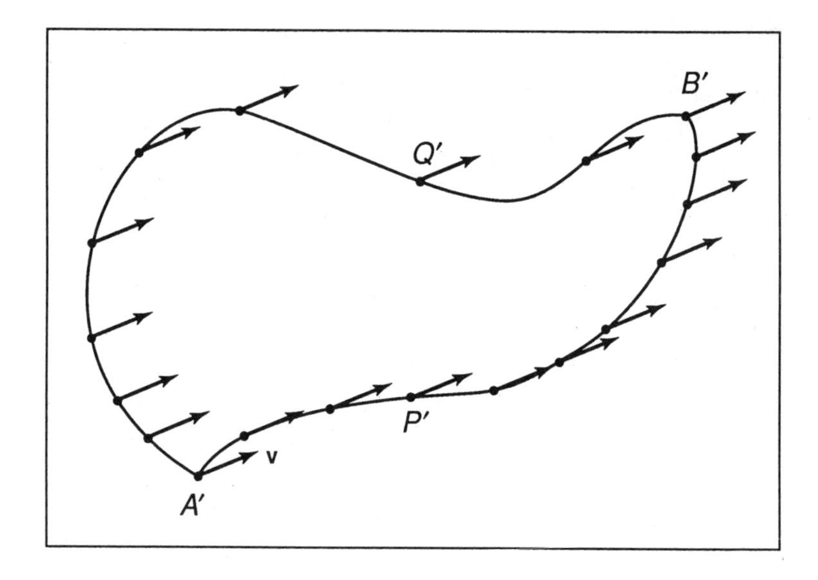

Figure 125 Parallel transport in the Euclidean plane

(Fig. 125). Pick an arbitrary vector **v** at A', and then at every point of the curve $A'B'$, draw a vector equal to **v**. This is always possible, since at each point of the curve, Euclidean parallelism guarantees the existence of a unique line having the direction of **v**, and from this line we can cut off a piece having the same length as the magnitude of **v**. Let us call this process *parallel transport* (or *parallel displacement* or *parallel translation)* of a vector along a curve in the plane. Notice that if a different plane curve $A'Q'B'$ is taken, and **v** is transported parallelly along it, the resulting vector at B' is the same as before. Also, if we transport **v** parallelly around the closed curve $A'P'B'Q'A'$, we end up with **v** again at the end of the circuit.

In 1917, the Italian mathematical physicist Tullio Levi-Civita extended the above notion to curved smooth surfaces in a most ingenious way, and thoroughly explored its analytical consequences (see his book: Levi-Civita (1926)). We may describe *parallel transport of a vector along a curve on a surface* in physical terms as follows:[1]

(i) Take any two distinct points A and B on a smooth, path-connected surface S and join them by any surface curve APB (Fig. 126). Let **v** be any vector at A which lies in the tangent plane to S at A.

(ii) Cut a narrow strip of paper of such a shape as to cover the curve AB, and trace AB on it. (Do not try to make the strip lie on the surface except along the curve, as this would generally require a non-isometric deformation of the strip.) Trace the vector **v** on the strip. (Squared paper works best here, as you can align one set of marked lines along **v**.)

(iii) Remove the strip from the surface and lay it flat on the plane. The space curve AB is thereby mapped into a plane curve $A'B'$ (Fig. 126b).

(iv) Transport the image of **v** on the strip parallelly from A' to B' (*i.e.*, use Euclidean parallelism here).

(v) Lay the strip back on the surface and secure it with tape. This will

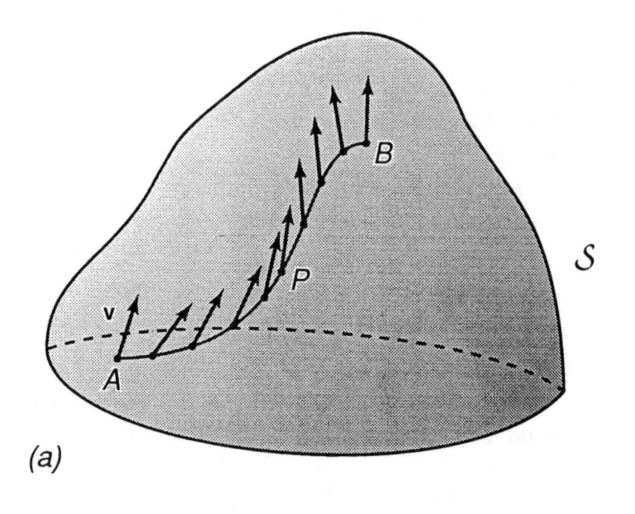

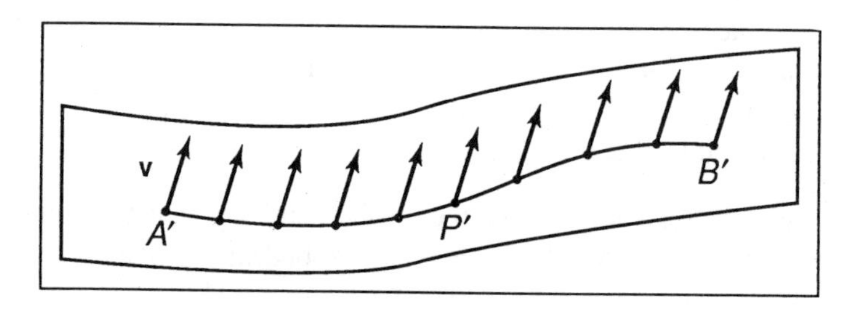

Figure 126 Parallel transport on a general surface

produce a set of vectors lying in tangent planes to the surface along AB.

Parallel transport is also called *Levi-Civita parallelism.* In extending this concept to surfaces with edges or vertices, let us agree not to lay the strip of paper across a vertex, and to cross the edges only transversally (*i.e.,* we shall always make a non-zero angle with an edge when we meet it).

Experiment 60 (Parallel transport): (**a**) Take the surface *EFGHIJ* which you prepared in Experiment 28, flatten it onto a plane, and join the points J and G by a straight line. Draw an arbitrary vector $\mathbf{v}$ at J and transport it parallelly along JG to G. Also transport $\mathbf{v}$ parallelly along *JIHG*. Place *EFGHIJ* on the intact box. The line joining J to G is now bent, and the vectors along the line now lie in two different planes. Argue that the vector which you would get by transporting $\mathbf{v}$ parallelly along an arbitrary path lying in the surface *EFGHIJ* and joining J to G, is independent of the choice of path.

(**b**) On the intact box, extend the line *EFG* around the back of the box and return to E, keeping away from the corners. Likewise, continue *JIH* around the box to close again at J. Transport $\mathbf{v}$ parallelly around any closed curve lying in the resulting strip. What do you find? □

Experiment 61 (Parallel transport of two vectors): Verify that if two vectors are transported parallelly along the same curve on a surface, then the angle between them remains constant. Hence, deduce that the inner product between the two vectors is also preserved. □

Experiment 62 (Parallel transport around a closed curve): On the surface *KLMNOPQ* of Fig. 84, transport a vector parallelly from P to M first via the path *PQLM* and then via the path *PONM*. What do you find? □

By now, you will have discovered the important fact that parallel transport of a vector is in general a *path-dependent* process. This geometrical phenomenon is called *holonomy*. To elaborate, if we join A and B in Fig. 126a by a curve other than APB, and transport $\mathbf{v}$ parallelly along the new curve, in general we will obtain a vector at B which has the same magnitude but a different direction than the original vector at B. Consequently, when we parallel-transport a vector once around a closed curve, in general we will not recover its original value. For example, on

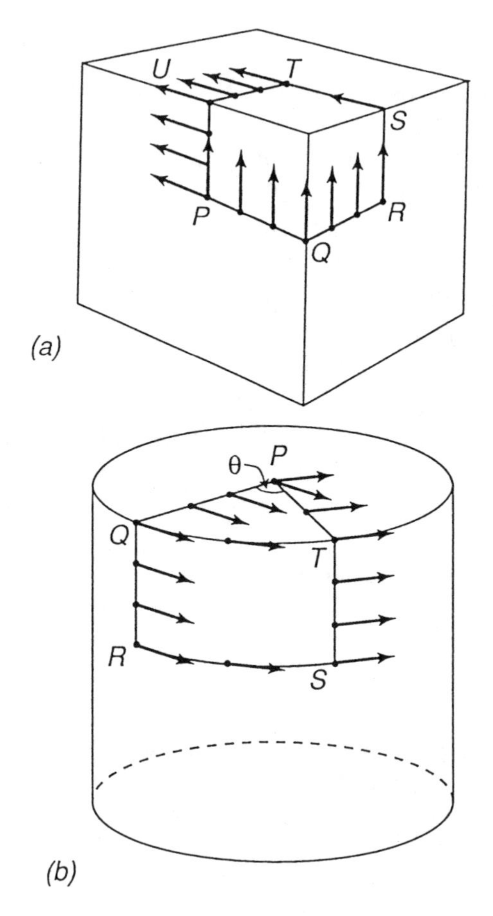

Figure 127 Change in direction of a vector upon parallel transport around a closed curve

the cube shown in Fig. 127a, a vector pointing vertically upwards at P is transported parallelly along PQR, and then along RST, and finally along TUP: it has rotated through 90 degrees counterclockwise in this process. As another example, consider the path $PQRSTP$ on the cylindrical container in Fig. 127b, where P is the center of the top of the container and QR and TS are two equal vertical line segments. Let θ be the measure of the angle QPT. Take a vector at P that lies in the top of the container and is perpendicular to the radius PQ. After its parallel transport around $PQRSTP$, the vector will be found to have rotated counterclockwise through an angle of measure θ.

We will call the angle through which a vector is rotated upon parallel transport around a closed curve the *holonomy angle* for the curve.

Experiment 63 (Holonomy angle): (**a**) For the box in Fig. 84, determine the holonomy angle that results from transporting a vector parallelly once around a closed curve surrounding (i) only two corners; (ii) only three corners; and (iii) only four corners.

(b) Parallel transport a vector around the closed path $RSTUVWR$ on the surface bounded by the lines $RSTUVWR$ and $\overline{R}\,\overline{S}\,\overline{T}\,\overline{U}\,\overline{V}\,\overline{W}\,\overline{R}$ in Fig. 84.

(c) Take the polyhedral surfaces you prepared in Experiment 29 and transport vectors parallelly around closed paths on them. Explore what happens when a circuit encloses a vertex, and then several vertices.

(c) Measure the holonomy angle for a circuit on an eggplant, watermelon, or large apple. □

Parallel Transport and Curvature

It should be clear to you by now that the property of local non-developability of a surface is intimately bound up with the path-dependency of Levi-Civita parallelism. Consider again the region Ω on the surface S in Fig. 119a. Let us transport a vector **v** parallelly once around the boundary of Ω, traveling in the counterclockwise direction (Fig. 128). Let the holonomy angle for this curve be ψ (measured in radians and considered positive in the counterclockwise direction). Then,[2]

$$\psi = \int_{\Omega} K \, da \; . \tag{19.1}$$

Hence, in view of Equations (16.1) and (16.4), the holonomy angle has the same value as the total curvature of the region Ω. This is a marvelous result in itself, and it also provides an easy method for measuring the total curvature of a region.

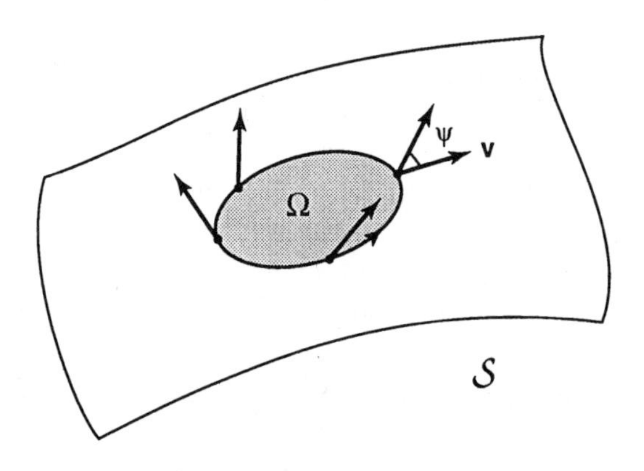

Figure 128 Parallel transport of a vector around the boundary of the region Ω

For each of the two objects in Fig. 121, the total curvature of the upper portion was found in Experiment 55: it has a value 2π. If you transport a tangent vector to *ABC* parallelly once around *ABC*, you will find that it rotates through a holonomy angle equal to 2π. Thus, equation (19.1) is verified for these objects.

If the region Ω has constant Gaussian curvature, Equation (19.1) reduces to

$$\psi = Ka \quad (K = constant) \; . \tag{19.2}$$

Thus, for a sphere of radius R,

$$\psi = \frac{a}{R^2} \, . \tag{19.3}$$

For the spherical segment considered in Example 1 of Chapter 16, $\psi = \theta \sin \phi$.

Experiment 64 (Holonomy angle on a sphere): (a) On a sphere, take two circuits, one of which surrounds the other. Transport a vector parallely around each circuit. Compare the resulting holonomy angles.

(b) Measure the holonomy angle for the path bounding the special spherical segment in Example 1 of Chapter 16.□

For a sufficiently small region Ω of area a,

$$\psi \approx K a \, . \tag{19.4}$$

We may use this result to study how the Gaussian curvature of a surface varies from point to point.

Experiment 65 (Intrinsic curvature of surfaces of objects): Take some containers, such as bowls, cans, or bottles, and a watermelon or eggplant, and transport vectors parallelly on their surfaces. Measure the holonomy angle for various circuits, and hence determine which parts of the surfaces have positive, or negative, or zero intrinsic curvature. □

In Figs. 129, 130, 131, and 132, photographs are reproduced of some common objects on which vectors have been transported parallelly. On the lateral surface of the cylinder in Fig. 129, the Gaussian and total curvatures are zero, and Levi-Civita parallelism is path-independent. If, however, a closed path is taken which includes a portion of the top of the container in addition to a portion of the lateral surface, the total curvature is non-zero and parallel transport is path-dependent (see Fig. 127b again). The total curvature of the cylinder is concentrated along the upper and lower rims. The Gaussian curvature tends to infinity as the rim becomes sharper and sharper.

In Fig. 130, vectors are shown on the outer surface of a mixing bowl. The bottom of the bowl is flat, and parallel transport of a vector from A along the two paths AMB and ANB yields the same vector at B. As one moves away from the bottom of the bowl, intrinsic curvature becomes apparent: notice the path-dependency of parallel transport around the closed path $DCABD$.

Fig. 131 shows a liquid-detergent bottle, on which closed paths have been taken (they extend slightly around the back of the bottle). The total curvature of the region enclosed by the lower circuit is positive, and a holonomy angle equal to about $7°$ was measured. On the other hand, the total curvature of the region enclosed by the upper circuit is negative, and a holonomy angle equal to about $-6°$ was measured.

For the balloon in Fig. 132, it is clear that the Gaussian curvature takes on positive, negative, and zero values, and correspondingly that parallel transport is not only path-dependent on the balloon, but that the degree of path-dependency varies as different closed paths are chosen.

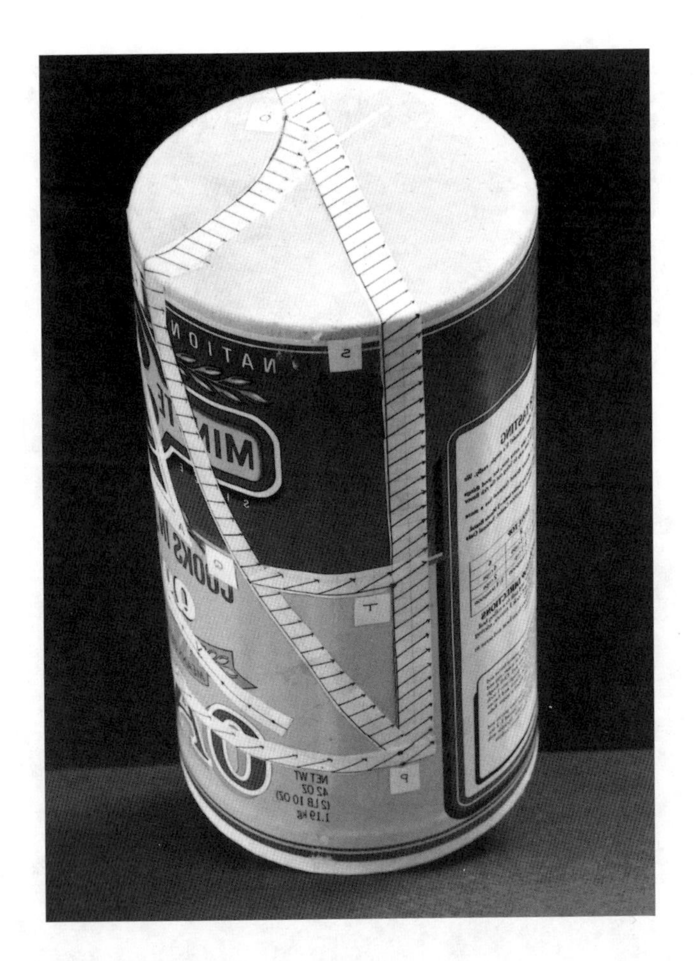

Figure 129 A cylindrical cardboard container

Figure 130 A mixing bowl

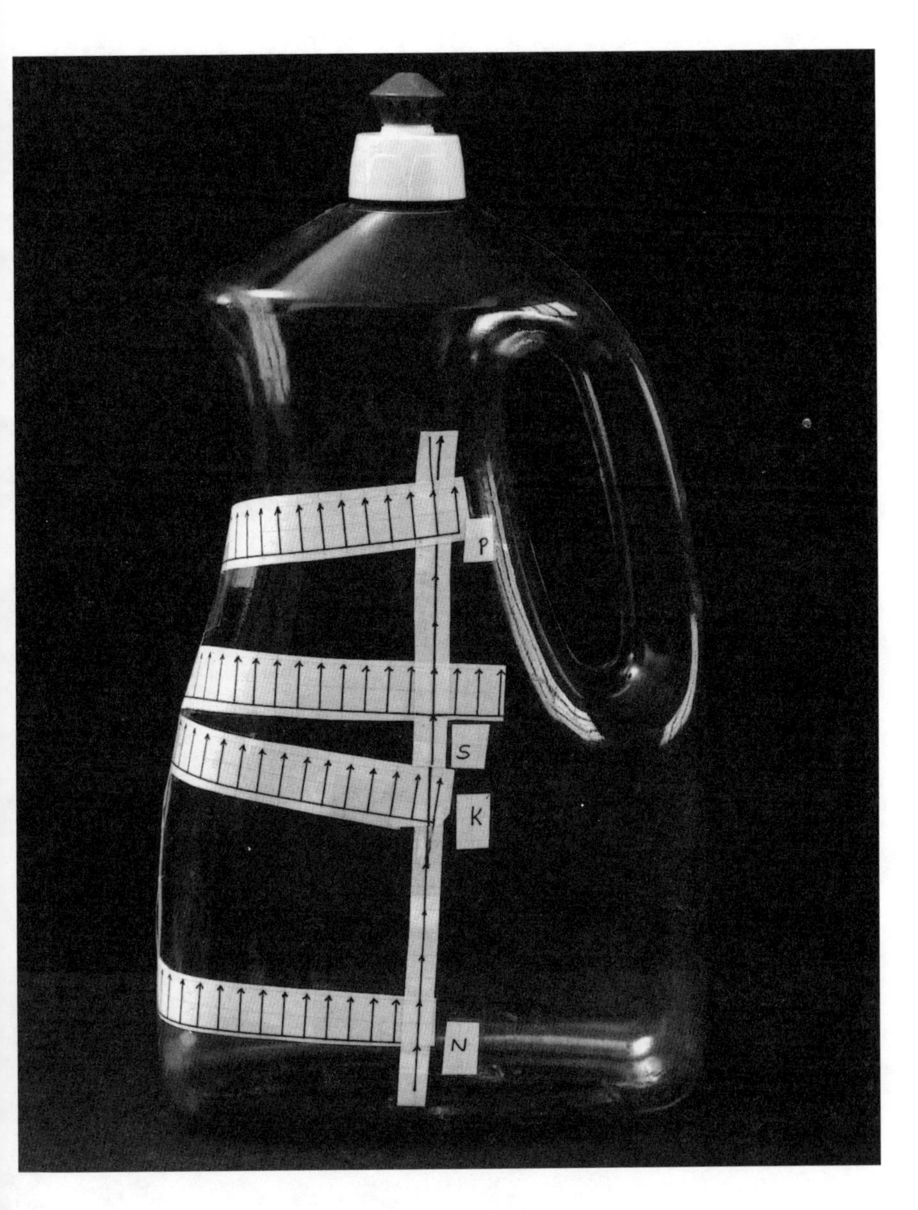

Figure 131 A liquid-detergent bottle

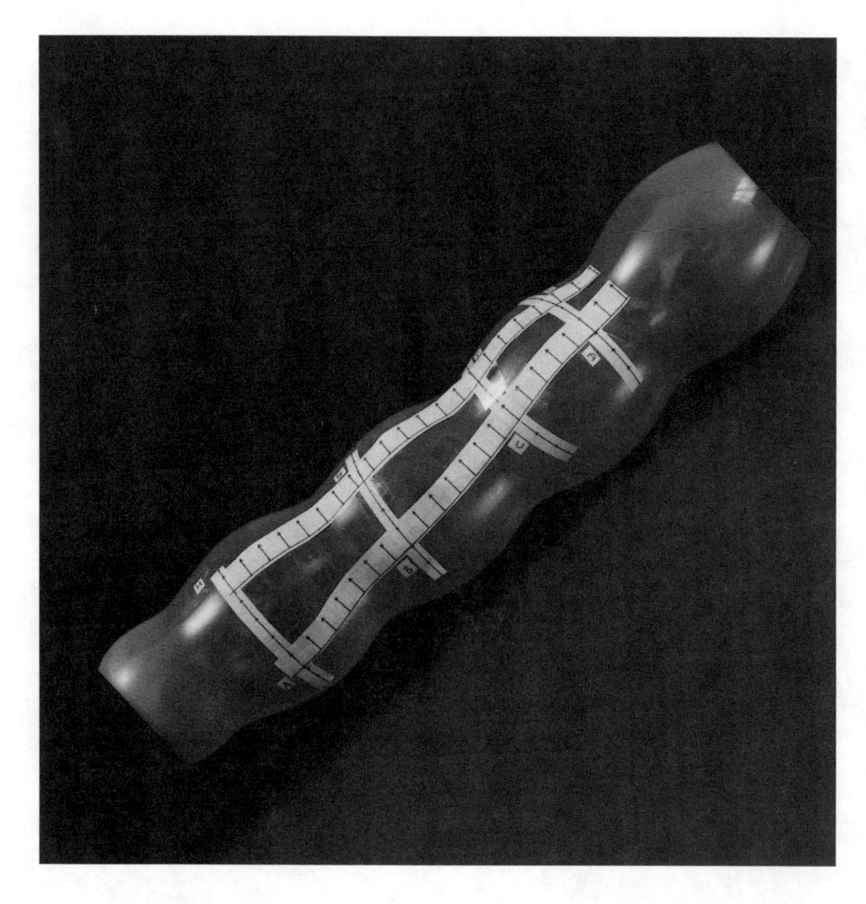

Figure 132 A balloon

Notes to Chapter 19

[1] Levi-Civita's own definition can be found in Sect. 10 of Ch. V of Levi-Civita (1926).

[2] See the theorem on pp. 390-391 of Vol. III of M. Spivak (1970). See also pp. 323-325 of O'Neill (1966).

20

Geodesics

In Euclidean plane geometry, the straight lines occupy a favored position among all of the curves that can be drawn in the plane. What are their analogues on a curved surface? One can try to generalize the property of *straightness*, or alternatively, one can utilize the other fundamental property of a straight line, namely that of being the *shortest* curve between two given points in the Euclidean plane.

Using the device of parallel transport, it is easy to extend the notion of straightness to curves on an arbitrary surface. Observe that straight lines in the Euclidean plane have the following property: Pick two arbitrary points A' and B' on a straight line (Fig. 133a). Draw a tangent vector to the line at A', and transport it parallelly to B'. Then, the resulting vector at B' is also tangent to the line. No other curves in the plane have this property. This suggests a definition: A "straight line", or *geodesic,* on a surface is a curve which possesses the property that the vectors obtained along it by parallel transport of a tangent vector constitute the tangent field itself ("autoparallelism"). Physically, we can obtain a portion of a geodesic on a surface by taking a narrow strip of paper on which a straight line has been drawn and laying it along the surface (Fig. 133b).[1] Some geodesics are shown on the objects in Figs. 129, 130, 131, and 132. It may also be noted that if a Euclidean straight line can be drawn on a surface, this line is automatically a geodesic. On a cylinder and a cone, one immediately recognizes one family of geodesics of this type. On a sphere, the geodesics are the great circles.

Experiment 66 (Geodesics): For the surfaces that you made in Experiments 28 and 29, construct geodesics through a chosen point A. How many can you draw through A? If B is another point on the same

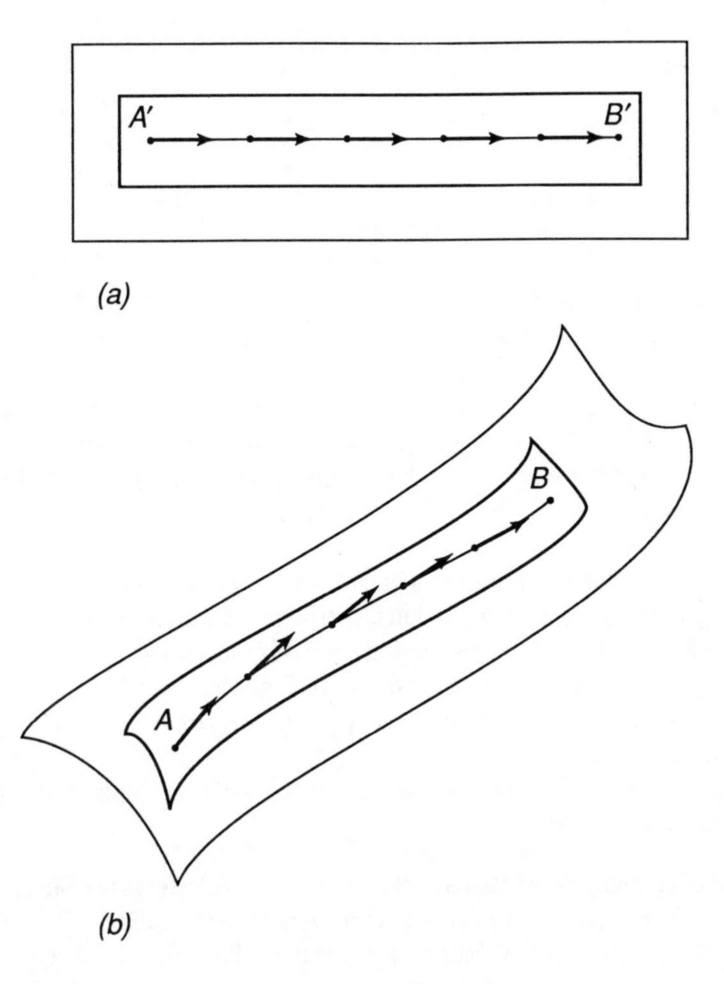

Figure 133 Constructing a geodesic on a surface

surface as A, can you always find a geodesic that joins A and B? Identify the different types of curves that form geodesics on a cylinder. □

Experiment 67 (Parallel transport along a geodesic): Demonstrate that if a vector is transported parallelly along a geodesic, then the angle that it makes with the tangent to the geodesic is preserved. □

Do geodesics also inherit the minimum-length property of straight lines? Here, one must be very careful. On a cylinder, it can be seen that two points A and B lying on the same generator but not on the same cross-section can be joined by at least two geodesics: a straight line and a helix. Likewise, two points on a sphere can be joined by the two complementary arcs of the same great circle. Thus, a geodesic is not necessarily the shortest curve between two given points. However, the following property does hold: every sufficiently small piece of a geodesic is the shortest curve between its end-points. It is important to note that the property of being a geodesic is an intrinsic property of a surface; in other words, a geodesic remains a geodesic under bending of the surface.

Experiment 68 (Lengths of geodesics): (a) Measure the lengths of the geodesics you constructed in Experiment 66 and compare them with the lengths of other curves joining pairs of points on geodesics.

(b) Cut a disk out of a sheet of paper. Find the shortest path between two points on the paper that lie on opposite sides of the hole. Is this path a geodesic?

(c) If a piece is removed from a ball, are shortest paths necessarily geodesics? □

Experiment 69 (Intrinsic character of geodesics): (a) Study how the geodesics on a cylinder behave as you bend the cylinder.

(b) On one piece of ball that you cut for Experiment 44, construct geodesics. Study how they behave as you bend the surface. □

For physically smooth convex surfaces, it is clear that geodesics can be obtained by pulling a string tautly between points on the surface. However, the frictional force between a real string and a real surface increases with increasing tension, and makes this method somewhat unsatisfactory.

The Geodesic Triangle

Consider a smooth surface $\mathcal{S}$ that is both arcwise connected and simply connected. A *geodesic triangle* on $\mathcal{S}$ is a figure formed by three geodesics meeting in three distinct points in the manner indicated in Fig. 134. Let

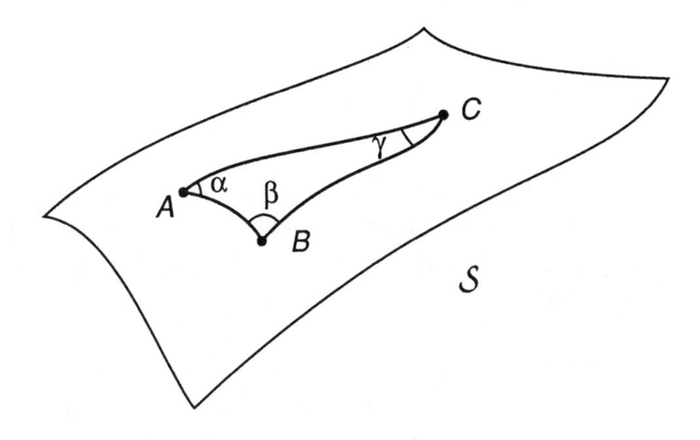

Figure 134 A geodesic triangle on a surface

the angles α, β, γ of the triangle be measured in radians, and let the region occupied by the triangle be denoted by Ω. Gauss proved that[2]

$$\alpha + \beta + \gamma - \pi = \int_{\Omega} K \, da \ . \tag{20.1}$$

Thus, the amount by which the angle-sum $\alpha + \beta + \gamma$ of a geodesic triangle algebraically exceeds π radians is equal to the total curvature of the geodesic triangle. This theorem, said Gauss[3],

> ... *if we mistake not, ought to be counted among the most elegant in the theory of curved surfaces* ...

For a small geodesic triangle,

$$\alpha + \beta + \gamma - \pi \approx K a , \tag{20.2}$$

where a is the area of the triangle. Recalling that in the Euclidean case

$$\alpha + \beta + \gamma = \pi , \tag{20.3}$$

we arrive at an interpretation of K as a measure of the deviation of a small geodesic triangle from a plane triangle. If $K > 0$, the sum of the angles in a geodesic triangle is greater than 180°, while if $K < 0$, the sum is less than 180°.

Experiment 70 (Laying special strips on a surface): (**a**) Draw one horizontal and two equal vertical lines and cut out a strip of paper in the shape indicated in Fig. 135, Try to form a geodesic triangle by laying this strip on the surfaces which you used in earlier experiments. Interpret your findings.

(b) Make other strips like the one in Fig. 135 but with different base angles, and try to form geodesic triangles with them on various surfaces. Measure the third angle in the geodesic triangle whenever you can. □

Experiment 71 (Geodesic triangle): Construct a geodesic triangle on a curved surface. Measure its angles. Starting at one vertex of it (*e.g.*, *A*

Figure 135 Preparing to make a geodesic triangle having 90° base angles

in Fig. 136) transport a vector parallelly around the triangle (the tangent vector to either *AC* or *AB* would be a good choice). Measure the angle through which the vector has been rotated. What do you conclude? □

In Experiment 70, one discovers a fascinating fact, namely that the amount by which the sum of the angles in a geodesic triangle exceeds 180° is equal to the holonomy angle for the triangle:

$$\psi = \alpha + \beta + \gamma - \pi \,. \tag{20.4}$$

The result (20.4) can be deduced immediately from equations (19.1) and (20.1). But, the following direct geometrical proof of it is very enlightening.

Proof: In a geodesic triangle ABC on a surface, let the vector **v** be tangent to *AB* at *A* (Fig. 136). Transport **v** parallelly along *AB*, and

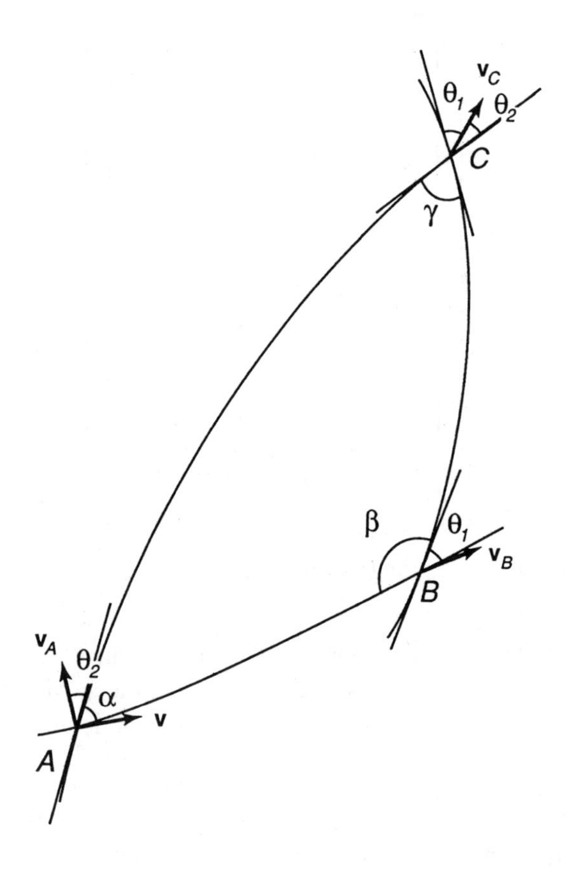

Figure 136 Parallel transport around a geodesic triangle

denote its value at B by $\mathbf{v}_B$. Since AB is a geodesic, $\mathbf{v}_B$ will be tangent to it at B. Now transport $\mathbf{v}_B$ parallelly along the geodesic BC, to arrive at $\mathbf{v}_C$ at C. Let θ_1 be the angle between $\mathbf{v}_B$ and the tangent to BC at B. Recalling the result stated in Experiment 61, we note that the angle which $\mathbf{v}_C$ makes with the tangent to BC at C will also be θ_1. Let the angle which $\mathbf{v}_C$ makes with the tangent to AC at C be θ_2. Transport $\mathbf{v}_C$ parallelly along the geodesic CA, to arrive at $\mathbf{v}_A$ at A. Then, the angle which $\mathbf{v}_A$ makes with the tangent to CA at A is also θ_2. Consequently, the value of the holonomy angle is

$$\begin{aligned} \psi &= \alpha + \theta_2 \\ &= \alpha + \gamma - \theta_1 \\ &= \alpha + \gamma - (\pi - \beta) \\ &= \alpha + \beta + \gamma - \pi \,. \end{aligned} \tag{20.5}$$

Experiment 72 (Holonomy around a geodesic triangle): Construct a geodesic triangle on a bowl, a vase, or a watermelon. Transport a vector parallelly around it and measure the holonomy angle. Also measure the interior angles of the triangle. Is the theoretical result (20.5) in agreement with your measurements? □

A convenient choice of geodesic triangle is one with 45°– base angles. You can form one by overlaying two narrow strips of paper of the type shown in Fig. 137. The strips are taped together along the base, but are left free at the vertex, forming a "flappy triangle".[4]

In Table 1 are listed the values of the holonomy angle for several surfaces.[5]

Table 1 Holonomy angle (in degrees) for various surfaces

Eggplant	55
Orange	46
Ball(5.25 – *inch diameter*)	30
Watermelon	22
Bowl	15
Sphere	5
Cone	0
Cylinder	0
Banana	– 42

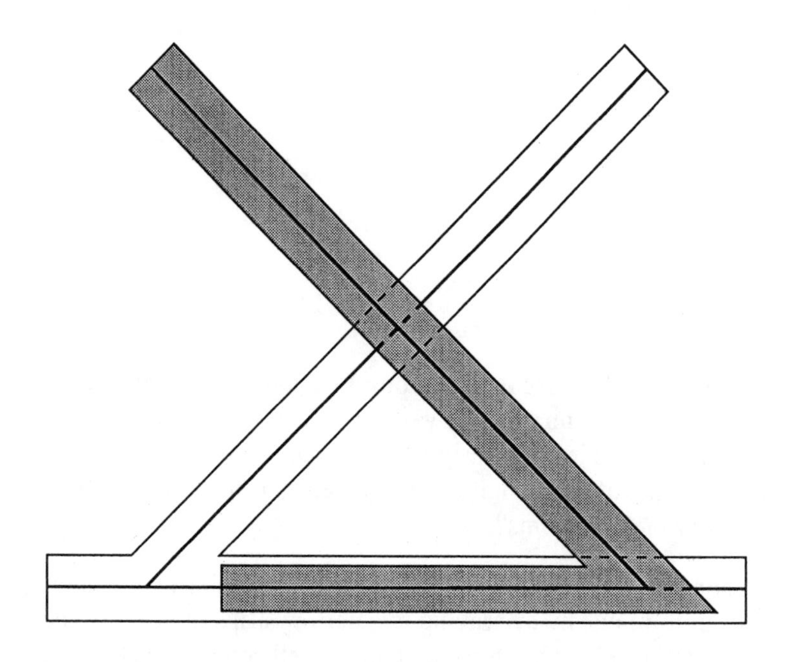

Figure 137 A flappy triangle

Notes to Chapter 20

[1] Beem (1976) also uses a paper-strip method for constructing geodesics, but he invokes a different geometrical property than parallel transport to justify the method.

[2] This result can be found on p. 30 of Gauss (1827). See also Section 4-8 of Struik (1961), where a more general result (the Gauss-Bonnet theorem) is proved.

[3] See p. 30 of Gauss (1827).

[4] This type of triangle was used in high-school classes by the author (see Casey (1994)).

[5] The values in Table 1 are extracted from the data given in Table 1 of Casey (1994).

21

Geometry and Reality

... it is certain that
mathematics generally,
and particularly geometry,
owes its existence
to the need which was felt
of learning something about
the behavior of real objects.

Einstein (1921)

As we saw in the preceding chapter, on a curved surface, geodesics replace the straight lines of the Euclidean plane. It also became clear that, in general, when figures are formed from geodesics, some of the most cherished results of Euclidean geometry – such as the theorem on angle sum of a triangle – must be given up. But, Euclid's theorems are derived by logical arguments from a set of postulates (or axioms). Physical intuition is allowed to enter the theory only through the postulates; after these are stated, only rules of logic can be appealed to in the proofs (although, of course, the inspiration for a theorem can come from intuition). It must therefore be the case that some of the properties ascribed to lines in Euclid's postulates cannot be true in general for geodesics. For, if all of Euclid's postulates held for geodesics, then so would his theorems. Let us examine some of the Euclidean postulates to see what goes wrong when they are interpreted to hold for geodesics.

Postulate 1: If A is any point and B is any other point, then there is one and only one straight line joining A to B.[1]

Another way of stating this postulate is that two points determine a straight line. Also, it follows from the postulate that two straight lines cannot enclose a space. Are these statements true for geodesics?

Experiment 73 (Postulate 1): Construct geodesics on a cylinder, a ball, and a vase, and test the validity of Postulate 1, both on a portion of each surface, and on the surface as a whole. State your conclusions. □

> *Postulate 2:* A finite straight line can be produced continuously in a straight line.

This postulate is interpreted to mean that the segment can be produced in a unique way from each end. Thus, two distinct straight lines could not have a common segment. What does the word "continuously" mean in Postulate 2? Presumably, it means that one can always produce the segment a little further, and hence that straight lines are endless. But, as Riemann realized (see p. 241), boundlessness does not imply infiniteness. A circle is of finite length, but it is endless; a fluid can flow within a torus without ever meeting an internal barrier. Euclid probably meant for all his straight lines to be producible beyond any finite length, although he did not explicitly say so.

Experiment 74 (Postulate 2): (a) Examine segments of geodesics on a cylinder and a ball to see whether they can be produced in accordance with Postulate 2.

(b) Are the geodesics on a sphere boundless? Are any of them infinite in length?

(c) On an infinite cylinder, are there both geodesics of finite and of infinite length. □

Postulates 3 and 4 of Euclid are about circles and right angles, respectively. Postulate 5, the Parallel Postulate, we have already met in Chapter 1 (see p. 20). An alternative to it, known as *Playfair's Axiom*, is:[2]

> *Postulate 5′*: Through any given point, and any given straight line not containing that point, only one parallel to the given line can be drawn.

Thus, in a plane, if AB is a given (infinite) straight line (Fig. 138) and P is any point that does not belong to AB, then there is not more than one straight line passing through P that is parallel to AB. Without any appeal

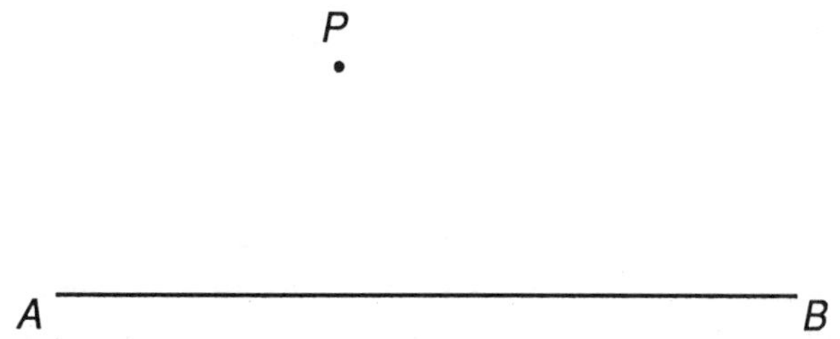

Figure 138 At most one Euclidean parallel to AB passes through P

to Postulate 5, Euclid is able to show that there does exist a straight line through P that is parallel to AB.[3] Consequently, in Euclidean geometry there is exactly one parallel to AB that passes through P.

Even in Greek times, some mathematicians felt uneasy about the Parallel Postulate, and for more than 2000 years attempts were made to prove it. Before studying the eventual resolution of this fundamental problem, let us explore the corresponding issue for geodesics. Thus, let P be a point on a surface S and let CD be a geodesic on S that does not pass through P. Then, does there always exist a geodesic which passes through P but never meets CD, and if so, is there only one such geodesic?

Experiment 75 (Intersecting and non-intersecting geodesics): Take three surfaces, for which $K = 0$, $K > 0$, and $K < 0$, respectively and tape a horizontal geodesic AB upon them. (A cylinder, a ball, and the liquid-detergent bottle in Fig. 131 are good candidates.) Pick a point P not lying on AB. Construct several geodesics through P and try to find one that does not intersect AB. □

In this, the last of our experiments, you will have discovered for yourself a crucial geometrical phenomenon: The existence and the uniqueness of a geodesic through P that does not intersect the given geodesic AB depend in an essential way on the intrinsic geometry of the surface.

In 1816, Gauss suggested privately that Euclid's Postulate 5 cannot be proved, but he never published his ideas on the subject. The credit of first presenting a non-Euclidean geometry must be given to the Russian geometer N.I. Lobachevsky (1792-1856), who on 23 February 1826 read a paper on the subject at The University of Kazan and later published his theory. Lobachevsky discovered that he could construct a logically consistent system of geometry in which Postulate 5 is negated, while Euclid's other assumptions are retained. Thus, he constructed a geometry in which through any point P not on a given line AB, at least two parallels to AB can be drawn. In this geometry, the sum of the angles of a triangle is less than 180°. Lobachevsky held the view that only physical observations could reveal whether the geometry of the real world is Euclidean or not. At virtually the same time as Lobachevsky, the Hungarian geometer János Bolyai (1802-1860) independently discovered the same non-Euclidean geometry as Lobachevsky.[4] The geometry of Lobachevsky and Bolyai is also called "hyperbolic geometry".

To appreciate the point of departure of the geometry of Lobachevsky and Bolyai from that of Euclid, consider again the point P and line AB in Fig. 138. From P, drop a perpendicular PM onto AB, and through P, construct a perpendicular PQ to PM (Fig. 139). Leaving Euclid's Postulate 5 aside, but retaining his other assumptions (both explicit and implicit), we can still deduce that PQ does not intersect AB – it is the Euclidean parallel.[5] Without Postulate V, we cannot say that PQ is the only straight line through P that does not intersect AB. Lobachevsky and Bolyai dared to imagine that there may be other straight lines, such as PE, that do not intersect AB no matter how far they are produced. Let us denote the angle MPE by ϕ. In Euclidean geometry, ϕ can have only the value 90°; in the absence of Postulate 5, ϕ may possibly also have a value less than 90°. Imagine that the non-intersecting line PE is rotated clockwise from a starting position PQ. As we do this, the angle ϕ decreases and PE passes continuously through a family of straight lines

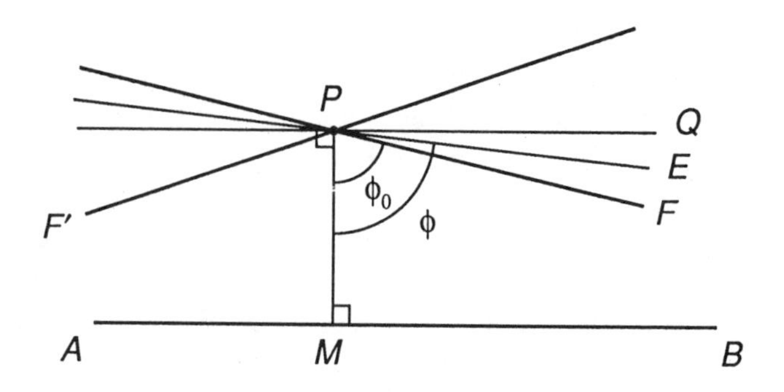

Figure 139 The Lobachevsky-Bolyai parallels PF and PF'

that do not intersect AB. Since PM intersects AB, there must be some value of ϕ that separates the non-intersecting lines from the intersecting ones. This value is called the *angle of parallelism*; let us denote it by ϕ_0. Every straight line for which $\phi \geq \phi_0$ never meets AB, and every straight line for which $\phi < \phi_0$ meets AB. The two straight lines PF and PF' are the *Lobachevsky-Bolyai parallels* to AB. In Euclidean geometry, ϕ_0 is forced to be 90°, and the two lines PF and PF' must coincide with PQ. If Postulate 5 is negated, ϕ_0 is less than 90° and the non-Euclidean geometry of Lobachevsky and Bolyai emerges.

It so happens that if the Gaussian curvature of a surface has a constant negative value, the local intrinsic geometry of the surface is the same as that of a piece of the Lobachevsky-Bolyai plane[6]. An example of such a surface is the funnel-shaped object shown in Fig. 140: it is called a pseudosphere.

The sphere, on the other hand, has constant positive Gaussian curvature, and on it all geodesics (*i.e.*, the great circles) intersect one another. The intrinsic geometry of a piece of the sphere corresponds to that of a piece of an alternative non-Euclidean plane ("elliptic geometry"). This geometry differs in an important way from the geometries of both Euclid, and Lobachevsky and Bolyai, in that its straight lines are finite in length. These lines are still unbounded (they do not end anywhere) and so must be closed curves. In elliptic geometry, Postulate 5 is replaced

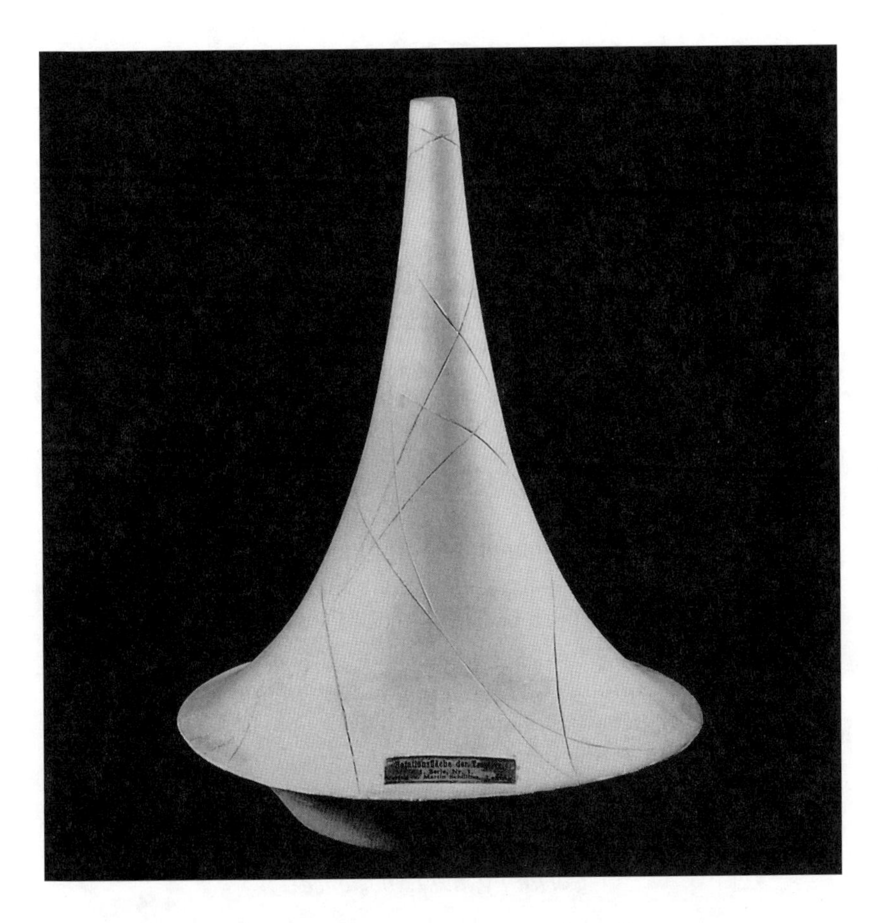

Figure 140 A pseudosphere [From Fischer (1986)]

by the assumption that no parallel to a given line can be drawn through a given point not lying on the line. Thus, in this geometry, all straight lines intersect. In elliptic geometry, the sum of the angles of a triangle is greater than 180°.

In this categorization of geometries, Euclidean geometry is known as "parabolic geometry".

On a general surface, the Gaussian curvature K can change from

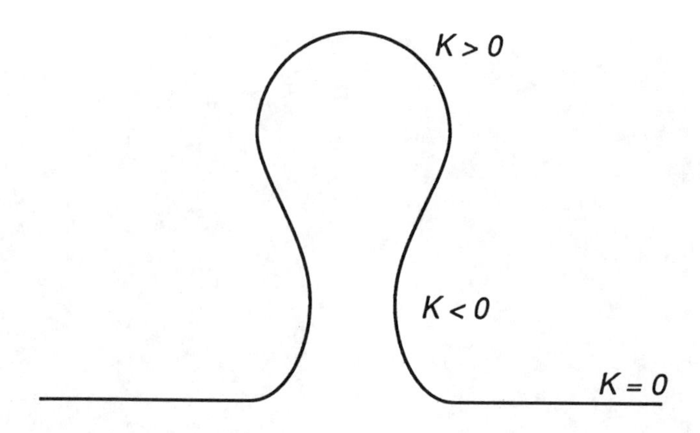

Figure 141 A bollard

point to point, and correspondingly, the local intrinsic geometry can change continually. For example, imagine that the Euclidean plane is deformed into a shape like a bollard (Fig. 141), with regions where K is constant. Then, the intrinsic geometry of the surface is elliptic at the top, hyperbolic at the middle, and parabolic at the bottom. More generally, K need not be constant over any piece of a surface: in that case, if K is positive at a point P, the local geometry is approximately elliptic in the vicinity of P, and similarly for the cases $K < 0$ and $K = 0$.

Certainty and Experience

The question of the relationship between mathematics and physical reality is one that has vexed philosophers and philosophically-minded mathematicians over the ages. The Pythagoreans held that mathematics was at the very core of reality (see p.10). Plato believed that the world of eternal ideas is the true reality, and that the physical world is just a poor reflection of it. From this viewpoint, Euclid's geometry would be the geometry of true reality, and the triangles, squares, and polyhedra that we might construct in physical space would be just approximations to the pure and eternal figures in the Platonic world of ideas. But, where then do the non-Euclidean geometries fit in? Surely, the eternal world could not admit contradictory geometries!

The resolution to this dilemma is the following: we must distinguish carefully between mathematical theories on the one hand, and sense-impressions and observations by means of scientific instruments on the other. As a result of man's experience with the natural world, primitive concepts and explanations arise. Typically, such explanations are inadequate in one way or another, and much revision has to take place. Eventually, a rather satisfactory theoretical structure, such as Euclid's geometry, may be developed. The theoretical structure, however, has an existence that is independent of what gave rise to it. The umbilical cord that ties it to reality can be cut. It is then a logical entity with its own internal rules. Theorems may be deduced, and calculations may be made without any reference to physical reality, if one so wishes.

Of course, if we compare the consequences of our deductions and calculations with empirical data, we may find small, or even large, disagreements. If the disagreement is small, we would say that the theory fits reality, whereas if the disagreement is large we would say that the theory is not at all an adequate description of reality. However, whether the theory fits reality or not, it still exists as a logical structure and may be very interesting as such. There are even theories which are so abstract that one would not even know how to begin to compare their results with reality, and yet these theories make fine mathematics.

It should be clear therefore that our experiences with the physical world can give rise to not just one theory, but many different theories, limited only by the power of human imagination. Moreover, it is to be expected that conflicting ideas and theories will arise, particularly in relation to the parts of our experience that are not yet accessible to precise measurement. Instead of a Platonic universe, what we have are the marvelous mental constructions of gifted theoreticians. The relationship between mathematics and reality was succinctly characterized by Einstein in the following way:[7]

> *as far as the propositions of mathematics refer to reality, they*

> *are not certain; and as far as they are certain, they do not refer to reality.*

But, by continuing to be guided by physical reality, mathematics can be assured of an inexhaustible source of inspiration.

Notes to Chapter 21

[1] The statements given here are based upon Euclid's literal statements and Heath's interpretations of them (see pp. 154-155 and pp. 195-220 of Vol. I of Heath (1926)).

[2]See Heath (1926), p. 220 and pp. 313-314.

[3]The existence of a parallel to AB through P is established in Proposition 27 of Book I of Euclid's *Elements* [Heath (1926)]. The proposition may be stated as follows: *If a straight line falling on two straight lines makes the alternate interior angles equal to one another, then the two straight lines are parallel to each other.* Its proof appeals to Proposition 16, which states: *If a side of a triangle is produced, the exterior angle is greater than each of the interior and opposite angles.* It is important to note that Proposition 16 holds only if every straight line is infinite in length, an assumption which is made only implicitly by Euclid.

[4] Translations of the work of Lobachevsky and Bolyai, as well as an extensive account of the historical development of non-Euclidean geometry, can be found in Bonola (1912). For additional information, see Manning (1901), Wolfe (1945), Coxeter (1947), Ch. XVII of Aleksandrov (1963), Gray (1979), and also the entries on Bolyai and Lobachevsky in the *Dictionary of Scientific Biography*.

[5] This follows from Proposition 27 (see Note 3).

[6] The Italian mathematician E. Beltrami (1835-1900) established this in 1868.

[7] This quotation and the one at the beginning of the chapter are taken from the brilliant lecture "Geometry and Experience" which Einstein gave before the Prussian Academy of Sciences on 27 January 1921. It can be found in Einstein (1954), pp. 227-240.

Bibliography

H. Abelson and **A. diSessa** (1980), *Turtle Geometry*. The MIT Press.

A.D. Aleksandrov (1963), Curves and Surfaces (Chap. III), and Non-Euclidean Geometry (Chap. XVII). In *Mathematics: Its Content, Methods, and Meaning,* 3 Vols. The MIT Press.

P.S. Aleksandrov (1963), Topology (Chap. XVIII). In *Mathematics: Its Content, Methods, and Meaning.* 3 Vols. The MIT Press.

F.J. Almgren, Jr. (1966), *Plateau's Problem.* W.A. Benjamin.

J.W. Armstrong (1976), *Elements of Mathematics,* 2nd edition. Macmillan Publishing Co.

B.H. Arnold (1962), *Intuitive Concepts in Elementary Topology.* Prentice-Hall.

S. Barr (1964), *Experiments in Topology*. T. Y. Cromwell Company, New York. Dover Publications, 1989.

J.K. Beem (1976), Measurement on Surfaces, *1976 Yearbook*: 156-162. National Council of Teachers of Mathematics.

E.T. Bell (1937), *Men of Mathematics*. Victor Gollancz, London. Simon and Schuster. Penguin Books, 1953 (2 Vols.).

R.H. Bing (1960), *Elementary Point Set Topology,* (The Eight Herbert Ellsworth Slaught Memorial Papers), *The American Mathematical Monthly*, **67** (No. 7, Supplement, August-September).

R. Bonola (1912), *Non-Euclidean Geometry*. The Open Court Publishing Company. Dover Publications, 1955.

C.B. Boyer (1949), *The History of the Calculus and its Conceptual Development [The Concepts of the Calculus]*. Hafner Publishing Company. Dover Publications, 1959.

C.V. Boys (1911), *Soap Bubbles*. Doubleday Anchor Books, 1959.

S. I. Brown and **M.I. Walter** (1990), *The Art of Problem Posing*, 2nd edition. Lawrence Erlbaum Associates, Publishers.

F. Cajori (1919), *A History of Mathematics*. The Macmillan Company. 4th edition, Chelsea Publishing Company, 1985.

J. Casey (1994), Using a Surface Triangle to Explore Curvature, *Mathematics Teacher*, **87**, 69-77.

S.-S. Chern (1990), What Is Geometry? *The American Mathematical Monthly,* **97**, 679-686.

W.K. Clifford (1882), *Mathematical Papers*. Macmillan & Co.

R. Courant (1934), *Differential and Integral Calculus*, Vol. I, 2nd edition, 1937 and Vol. II, 1936. Interscience Publishers (John Wiley & Sons).

R. Courant (1940) Soap Film Experiments with Minimal Surfaces, *The American Mathematical Monthly,* **47** , 167-174.

R. Courant and **H. Robbins** (1941), *What is Mathematics?* 4th edition, 1947. Oxford University Press.

J.L. Coolidge (1940), *A History of Geometrical Methods*. Oxford University Press. Dover Publications, 1963.

H.S.M. Coxeter (1947), *Non-Euclidean Geometry*, 2nd edition. The University of Toronto Press.

H.M. Cundy and **A.P. Rollett** (1961), *Mathematical Models,* 2nd edition. Oxford University Press.

Dictionary of the History of Ideas (1973), P.P. Wiener (Editor in Chief), 4 Vols. + Index. Charles Scribner's Sons.

Dictionary of Scientific Biography (1970), C. Coulston Gillispie (Editor in Chief), 18 Vols. Charles Scribner's Sons.

R. Dugas (1955), *A History of Mechanics*. Éditions du Griffon, Neuchatel. Dover Publications, 1988.

W. Dunham (1990), *Journey through Genius*. John Wiley & Sons. Penguin Books, 1991.

G. Waldo Dunnington (1955), *Carl Friedrich Gauss: Titan of Science*. Hafner Publishing Co.

A. Einstein (1954), *Ideas and Opinions*. Dell Publishing Co.

The Encyclopedia of Philosophy (1967), P. Edwards (Editor in Chief), 8 Vols. Macmillan.

H. Eves (1990), *An Introduction to the History of Mathematics,* 6th edition. Saunders College Publishing.

B. Farrington (1953), *Greek Science*. Pelican Books.

J. Fauvel, R. Flood, and **R. Wilson**, Editors (1993), *Möbius and his Band.* Oxford University Press.

G. Fischer (1986), *Mathematische Modelle - Mathematical Models*, and *Commentary*. F. Vieweg & Sohn, Braunschweig/Wiesbaden.

C. Folio (1985), Bringing Non-Euclidean Geometry Down to Earth, *Mathematics Teacher,* **78**, 430-431.

A.T. Fomenko and **A.A. Tuzhilin** (1991), *Elements of the Geometry and Topology of Minimal Surfaces in Three-Dimensional Space. Translations of Mathematical Monographs*, **93.** American Mathematical Society.

H. Freudenthal (1962), The Main Trends in the Foundations of Geometry in the 19th century, in *Logic, Methodology and Philosophy of Science*, pp. 613-621. Stanford University Press.

K.F. Gauss (1827), *General Investigations of Curved Surfaces*, Translated by A. Hiltebeitel and J. Morehead. Raven Press Books, 1965.

J. Gray (1979), *Ideas of Space*, 2nd edition, 1989. Oxford University Press.

V.L. Hansen (1993), *Geometry in Nature.* A.K. Peters, Wellesley, Massachusetts.

T.L. Heath (1912), *The Works of Archimedes.* Cambridge University Press. Dover Publications.

T.L. Heath (1921), *A History of Greek Mathematics,* 2 Vols. Oxford University Press.

T.L. Heath (1926), *The Thirteen Books of Euclid's Elements,* 3 Vols., 2nd edition. Cambridge University Press. Dover Publications, 1956.

D. W. Henderson (1996), *Experiencing Geometry on Plane and Sphere.* Prentice Hall.

L.D. Henderson (1983), *The Fourth Dimension and Non-Euclidean Geometry in Modern Art.* Princeton University Press.

R. Hermann (1975), *Ricci and Levi-Civita's Tensor Analysis Paper.* Math Sci. Press.

D. Hilbert and **S. Cohn-Vossen** (1952), *Geometry and the Imagination.* Chelsea Publishing Company.

P. Hilton and **J. Pedersen** (1988), *Build your own Polyhedra.* Addison Wesley Publishing Company.

S. Hollingdale (1989), *Makers of Mathematics.* Penguin Books.

G.S. Kirk and **J.E. Raven** (1957), *The Presocratic Philosophers.* Cambridge University Press.

C. Lanczos (1970), *Space Through the Ages.* Academic Press.

Lao-Tzu, *Tao Te Ching: The Canon of Reason and Virtue*, translated by D.T. Suzuki and P. Carus. Open Court, 1974.

H. Lebesgue (1966), *Measure and the Integral.* Holden-Day.

P. Le Corbeiller (1954), The Curvature of Space, *Scientific American*, **191**, 80-86 (November issue).

T. Levi-Civita (1926), *The Absolute Differential Calculus*. Blackie & Son Limited, London and Glasgow. Dover Publications, 1977.

T. Levi-Civita (1954), *Opere Mathematiche*, 4 Vols. Nicola Zanichelli Editore, Bologna.

D.R. Lichtenberg (1988), Pyramids, Prisms, Antiprisms, and Deltahedra, *Mathematics Teacher*, **81**, 261-265.

Sir Henry Lyons (1927), Ancient Surveying Instruments, *The Geographical Journal*, **69**, 132-143.

M. P. Lipschutz (1969), *Differential Geometry*, Schaum's Outline Series. McGraw-Hill Book Company.

H. P. Manning (1901), *Introductory Non-Euclidean Geometry.* Ginn and Company. Dover Publications, 1963.

J. McCleary (1994), *Geometry from a Differentiable Viewpoint.* Cambridge University Press.

J. Clerk Maxwell (1873), *A Treatise on Electricity and Magnetism*, 3rd Edition. Clarendon Press, 1891. Dover Publications, 1954, 2 Vols.

A.F. Möbius (1886), *Gesammelte Werke,* edited by F. Klein. Verlag von S. Hirzel, Leipzig.

M. Monastyrsky (1987), *Riemann, Topology, and Physics*. Birkhäuser.

B. Moses, E. Bjork, and **E.P. Goldenberg** (1990), Beyond Problem Solving: Problem Posing, *1990 Yearbook: Teaching and Learning Mathematics in the 1990s*: 82-91. National Council of Teachers of Mathematics.

J. Muir (1961), *Of Men and Numbers*. Dell Publishing Co.

B. O'Neill (1966), *Elementary Differential Geometry*. Academic Press.

O. Neugebauer (1957), *The Exact Sciences in Antiquity,* 2nd. edition. Brown University Press. Dover Publications, 1969.

R. Osserman (1986), *A Survey of Minimal Surfaces*. Dover Publications.

R. Osserman (1990), Curvature in the Eighties, *The American Mathematical Monthly,* **97**, 731-756.

R. Osserman (1995), *Poetry of the Universe*. Anchor Books, Doubleday.

R. Penrose (1980), The Geometry of the Universe, in *Mathematics Today*: 83-125. Vintage Books, New York.

A. Pugh (1976), *Polyhedra: a Visual Approach*. University of California Press.

B. Riemann (1876), *Gesammelte Mathematische Werke*. Teubner (1876; 2nd edition, 1892), Dover Publications (1953), Springer-Verlag (1990).

Satya Prakash Sarasvati (1987), *Geometry in Ancient India*. Govindram Hasanand, Delhi.

W. Sartorius von Waltershausen (1856), *Gauss,* [*A Memorial*], S . Hirzel, Leipzig. Translated into English by Helen Worthington Gauss, Colorado Springs, Colorado (1968).

A. Seidenberg (1962a), The Ritual Origin of Geometry, *Archive for History of Exact Sciences*, **1**, 488-527.

A. Seidenberg (1962b), The Ritual Origin of Counting, *Archive for History of Exact Sciences*, **1**, 1- 40.

A. Seidenberg (1978), The Origin of Mathematics, *Archive for History of Exact Sciences*, **18**, 301-342.

A. Seidenberg (1981), The Ritual Origin of the Circle and the Square, *Archive for History of Exact Sciences*, **25**, 269-327.

A. Seidenberg (1983), The Geometry of the Vedic Rituals, in F. Staal (ed.), *Agni, the Vedic Ritual of the Fire Altar*, 2 Vols., Asian Humanities Press, Berkeley, Vol. 1, pp. 95-126.

S.N. Sen and **A.K. Bag** (1983), *The Śulvasūtras* . The Indian Natural Science Academy, New Delhi.

G. F. Simmons (1992), *Calculus Gems*. McGraw-Hill.

D. E. Smith (1929), *A Source Book in Mathematics*. Dover Publications, 2 Vols., 1959.

D. E. Smith and **M. L. Latham** (1954), *The Geometry of René Descartes,* Dover Publications.

W.G. Spencer (1876), *Inventional Geometry*. American Book Company.

M. Spivak (1970), *A Comprehensive Introduction to Differential Geometry*, 5 Vols. Publish or Perish.

D.J. Struik (1961), *Lectures on Classical Differential Geometry,* 2nd edition. Addison-Wesley. Dover Publications, 1988.

I. Thomas (1951), *Selections Illustrating the History of Greek Mathematics*, 2 Vols. Harvard University Press, Cambridge, Massachusetts, and W. Heinemann Ltd., London.

B. L. van der Waerden (1961), *Science Awakening*. Oxford University Press.

B. L. van der Waerden (1983), *Geometry and Algebra in Ancient Civilizations*. Springer-Verlag.

M.J. Wenninger (1966), *Polyhedron Models for the Classroom*, Reston, Va. National Council of Teachers of Mathematics.

M.J. Wenninger (1971), *Polyhedron Models*. Cambridge University Press.

H. Weyl (1922), *Space - Time - Matter*. Dover Publications.

A.N. Whitehead (1958), *An Introduction to Mathematics*. Oxford University Press.

G.T. Whyburn (1942), What is a Curve? *The American Mathematical Monthly,* **49**, 493-497.

H.E. Wolfe (1945), *Introduction to Non-Euclidean Geometry*. The Dryden Press, New York.

E.L. Woodward (1990), High School Geometry Should be a Laboratory Course, *Mathematics Teacher*, **83**, 4-5.